Wild flowers of the countryside

HILARY BLADON

A Look-out Chart Handbook
Series Editor: Treld Pelkey Bicknell

Colour Illustrations: Harry Titcombe
Black and White Illustrations: Patti Pearce

Concertina Publications Limited
Covent Garden, London

First published in Great Britain by
Concertina Publications Limited
11–13 Broad Court, Covent Garden,
London, W.C.2, 1979

ISBN 0 905916 10 7

Printed by Butler & Tanner Ltd, Frome and London

For Lauren and Meriel

Contents

HILARY BLADON grew up in urban Liverpool but nonetheless developed an absorbed and lasting passion for natural history. She studied botany and zoology at the Royal Holloway College of London University gaining not only a first-class honours degree but also a prize for outstanding merit. Married with two small daughters she now lives on the 'windiest hilltop in Dorset' surrounded by, and writing about, the life of the countryside.

PATTI PEARCE trained at Winchester School of Art and London University Institute of Education. She went to Kenya after her training and taught there for 7 years, travelling widely and exhibiting her work in Nairobi. She now lives in Buckinghamshire where she is head of Art Department at the Royal Latin School in Buckingham. Her interests include photography and painting and she regularly exhibits her work.

Introduction

PLANTS HAVE LIVED ON OUR PLANET EARTH FOR 2,000 million years, and today more than half of those living (as many as 250,000 species in fact) are classified as flowering plants, or *angiosperms* – that is, plants with enclosed seeds. The success of the angiosperms is largely due to their unique structural feat, the flower.

How much pleasure wild flowers, in all their varied forms, give us; their beauty has inspired mankind for centuries. The gorgeous hues of exotic flowers such as the sacred LOTUS (*Nelumbium*) of the Indian rivers with its huge blooms, or the intricate orchids of tropical forests, are certainly beautiful, and the bizarre forms of Mexican cacti quite fascinating; yet the humble plants of our own wayside, meadow or woodland offer us much beauty of form and colour too. Consider the delights of a walk through a wood in April: easy to spot are the bright yellow flowers of the PRIMROSE (*Primula vulgaris*), massed together on their crumpled beds of soft, wrinkled leaves, and the WOOD ANEMONE (*Anemone nemorosa*), with its white or pale mauve flowers drooping over delicately divided leaves. The LESSER CELANDINE (*Ranunculus ficaria*) has brightly coloured flowers, like burnished gold stars shining from amongst its heart-shaped leaves, and the SWEET VIOLET (*Viola odorata*) may be spotted by its beautiful purplish-blue flowers, with their delightful fragrance. Later there will be PURPLE ORCHIS (*Orchis mascula*), and carpets of BLUEBELLS (*Endymion non-scriptus*), to delight the eye. Less easy to find are the tiny green- or brown-coloured flowers of springtime such as those of the early-flowering

SWEET VERNAL-GRASS (*Anthoxanthum odoratum*), bearing its flowers of leaf-green where it grows beneath the oaks, themselves decked in green catkins of flowers at this time of the year. Forming the background to the scene are the flowerless plants, the clumps of moss and little, tightly furled bracken fronds.

Weeds too, so readily dismissed because they grow, by definition, where they are not wanted, have a beauty of their own like the tissue-flimsy petals of the summer-flowering FIELD POPPY (*Papaver rhoeas*), brilliant scarlet amongst the golden corn, or the massed florets of the little DAISY (*Bellis perennis*) flower, so common on our lawns that it is sometimes called 'Mother of Thousands'. The SCARLET PIMPERNEL (*Anagallis arvensis*), though so tiny, has little blossoms of such an intense red that they are an easy and delightful find as soon as the corn harvest is cut.

The flower, then, has obvious attractions in itself. However, plants do not produce glorious flowers to satisfy man's aesthetic sense, but to ensure the continuation of their species, for all our more conspicuous flowers are only leaves modified in order to attract animal, usually insect, pollinators. But we must not forget that, as a reproductive aid, the flower is a temporary structure, and that behind and beneath lies the vital 'back-up system', the basis of the plant, without which the flower could not operate, and fascinating in itself. For plants, with their green pigment chlorophyll, can create the complex materials which are the source of the plant's own energy and, too, the source of the world's host of living organisms, man included, through their utilization of the energy of sunlight.

Consider too the diversity amongst flowering plants. Woody or herbaceous, they range from majestically towering trees which can flower every year over a period of many hundreds of years, to lowly herbs that creep the ground like the GROUND IVY (*Glechoma hederacea*), and little annuals like the scarlet pimpernel, which over-winter as seeds, with a multitude in between; shrubs like the WHORTLEBERRY (*Vaccinium myrtillus*); herbaceous perennials (surviving for several years) like the ROSEBAY WILLOWHERB (*Epilobium angustifolium*);

plants which over-winter as bulbs like the SNOWDROP (*Galanthus nivalis*), as well as marsh and water plants.

Leaves can be simple or composed of several leaflets, and vary from the upright green swords of the YELLOW FLAG (*Iris pseudocorus*), to the heart-shaped blades of the sweet violet, from the finely toothed as in many BUTTERCUPS (*Ranunculus* species), to the needle shapes of the LADY'S BEDSTRAW (*Galium verum*).

In fact the variation possible is almost endless; there are as many different forms of flowering plant as there are niches for them to occupy, for the angiosperms are not only the highest forms of plant life but the most diversified and widespread, and can be found growing almost everywhere on the earth – in water, in deserts, on mountains and in different kinds of soil. Plants have adapted to suit the places in which they grow by the process of evolution, by which, generation after generation, natural selection takes place. Amongst varied offspring, those better adapted to suit the environment tend to survive and produce more offspring with the advantageous variations. At the time the great forests were forming the coal deposits, the flowering plants did not even exist; they are in fact a group which has evolved relatively recently, and their success has gone hand in hand with a decline of earlier plant groups like the ferns, clubmosses and horsetails. Seed-bearing plants, then, are the height of plant development; they occupy the top two divisions of a total of seventeen which span the entire plant kingdom from trees to microscopic scum floating on the surface of pondwater.

The first and most primitive of these groups is the *gymnosperms*, so-called because of their 'naked seeds', and it is in this division that conifers (larch, pine, cedar, etc.) belong. Angiosperms, the higher flowering plants with their enclosed seeds, occupy only the last division. They themselves are divided into two groups based upon a difference in number of their seedling leaves, the *cotyledons*. The *dicotyledons*, the more primitive group and covering most of our net-veined plant species, possess two of these seedling leaves, raised like two tiny green ping-pong bats upon germination of the seed, whilst the *monocotyledons* have but one. The latter group, which includes the grasses, has members which are characterized by their parallel-veined leaves and flower parts arranged in threes or multiples of three.

A classification system of this sort is, of course, extremely necessary. We have noted that great variety exists amongst the flowering plants, but when faced with the diversity in the plant kingdom as a whole, it becomes a truly bewildering task to speak efficiently of them without the use of some widely accepted, workable system of classification. All need to be seen against a background of their evolutionary relationships.

The basic format and super-imposed adaptations, in all their vast array, shown by the flowering plants of today, are surely miracles of effectiveness. This is real necessity, for there is always another plant waiting to take advantage of an unusual ecological niche, in addition to the competition that exists for the young plant amongst its fellow-seedlings for limited resources. (If all the germinating seedlings from a single MULLEIN PLANT (*Verbascum thapsus*) did survive, they would cover an area the size of a football pitch!) But despite this element of constant competition, individual species do not live in splendid isolation; delicate inter-relations exist and together they form the environment of all living things, like a huge infinitely complex web of interconnecting strands. A strand broken in one place can have far-reaching consequences....

Despite the great richness of the plant kingdom, a terrifying 10% of the world's flora is believed to be on the brink of extinction or dangerously rare, as a result of man's activities. Man's relations with plants do not stop with the pleasure afforded by the beauty of wild flowers. He has total dependence on the plant kingdom for his basic needs of food and oxygen, as do all other animals, and in addition they provide man with timber, various drugs and other important products. Ties that have linked man and nature through the ages are being broken. It is time we reawakened our awareness of these ties before it is too late.

1: Plants and how they grow

WHATEVER THE PECULIARITIES OF THEIR STRUCTURAL adaptation, at 'grass roots' level plants are all the same. As with all living things, they are built of cells. Angiosperms have as many as seventy or eighty different kinds of cells arranged in, and capable of organization within, tissues which provide a framework for the plant, its water and food supply, storage mechanisms and protective coverings. All the various kinds of plants are composed of these basic units, following their individual internal guidelines that will determine whether the outcome is an oak or a mushroom, a pine needle or a blade of grass. In order to grow, plants need light, water, warmth, air and nutrients. It is the difference in availability of these essentials which varies the ecological niches available to plants, and hence determine the character of the plants associated with that niche.

The higher plant consists of four basic elements – the leaf, stem, roots and flower. The leaf is the 'powerhouse' of the plant, the prime site of food manufacture. Its cells, in common with the cells of all green plants, contain within them a unique structure able to control and contain the amazing chemical reaction whereby plants can trap and use light energy from the sun. These structures are called *chloroplasts*, and the process by which this light energy is trapped and utilized to make energy-storing carbohydrate from gaseous carbon dioxide and water, is called *photosynthesis*. The process releases oxygen into the atmosphere; bubbles of oxygen can often be seen clinging to the leaves of submerged water plants. The special light-trapping substance is arranged in layers within the chloroplast, acting as a receiving array for the light energy. The substance, a catalyst for the chemical reaction of photosynthesis, is the green pigment chlorophyll; plants appear green because this is the only part of 'white' light that cannot be used by the plant so it is reflected back whilst the other colours are absorbed. After it is trapped, the light energy is stored as 'high-energy substances' and these are called upon when energy is needed to drive the complex series of reactions necessary in the formation of sugars and starches that plants need for growth.

This food is transported from the chloroplasts to wherever it is needed, to build basic structural parts, to be stored as food reserves for seeds, or to provide energy in the many complex processes entailed in ordinary growth. This release of energy from the carbohydrates is termed *respiration* and, again, there are special sites within the cell for this activity, namely the *mitochondria*. Essentially the process is a reversal of photosynthesis, oxygen being required and carbon dioxide and water being the by-products of the process.

Gaseous exchange between the leaf and the atmosphere is obviously essential if these processes are to take place and so we find pores (*stomata*) on the leaf surface, usually the lower. Water-lilies, with their floating leaves, have to have them on the upper surface. Inside the leaf there are large spaces near the pores so that a large cell surface area is presented for the exchange of gases. Some control, too, is necessary to compensate for changes in the weather; this is achieved by the ability of stomata to open or close according to conditions. Many species also show adaptations in their leaf form to counteract excessive water loss from the plant through the stomata by positioning their pores in pits or having leaves tightly rolled with the pores only on the inside as in MARRAM GRASS (*Ammophila arenaria*).

Sunlight, too, is necessary for photosynthesis, so close-packed trees in a forest grow tall with few lower branches to try and outstrip their neighbours in the battle for sunlight. Other plants have their stem or leaves so arranged that the maximum leaf area is presented to the sunlight, as in the spiral pattern of leaf arrangement shown by many plants, a design which maximizes exposure to the sunlight and minimizes shading by the other leaves on the plant. Leaves themselves are usually thin and flat to present a large surface area to absorb light energy though, again, many modifications exist. These are chiefly to do with minimizing water loss as in the needle-shaped leaves of the conifers which are adapted to survive a period when the ground is frozen, or the fat, water-storing, waxy leaves of desert-growing succulents. The aptly named COMPASS PLANT (*Lactuca serriola*), a common wayside plant of Europe and America, holds its stem leaves upright in a north–south plane to prevent overheating.

Every year photosynthesis is responsible for the production of 20 cubic kilometres (5 cubic miles) of vegetation on our earth, enough to fill twenty thousand million cubes, each with sides one metre long. In our oceans, this amount is almost four times greater – an important process indeed! Yet healthy plants need more than just the carbohydrates that the green aerial parts can manufacture, for their bodies are composed of many other elements, over fifty in the case of some plants. The most important elements generally are nitrogen, phosphorus, potassium, calcium, magnesium, sulphur and iron; other elements are of varying importance to plants. The HORSETAIL (*Equisetum*), for instance, contains relatively high proportions of silicon, whilst some species of MILK-VETCH (*Astragalus*) have been found to contain selenium.

Without water too, plant life, as with the rest of living things, cannot exist on earth; the true desert wastes are barren. Life first began in it, and still today is ultimately dependent upon it. Though the body of a living plant may contain between 50% and 98% water, only 0.2% is used directly in the plant's body-building processes. Of the rest, a proportion maintains the plant's rigidity, in other words, stops it wilting, while most simply evaporates into the air, a process known as transpiration. A typical plant of our lattitudes would use 200 to 500 grams (0.35–0.87 pints) of water with dissolved nutrients to make one gram (0.35 ounces) of solid tissue.

Though some plants such as CLOVER (*Trifolium repens*) have special root nodules whereby they can extract nitrogen from the air, plants acquire their nutrients along with water chiefly from the soil. They are able to do this by the development of an extensive (usually) root system, which not only serves to anchor the plant in the ground and function sometimes as a food storage organ, but brings the millions of unicellular root hairs into contact with as great an area as possible of soil water and mineral solutions, where they act as a sort of pumping device. Beneath the soil, a plant's roots will occupy at least as much space as its aerial parts for they are inter-dependent; the shoots need the nutrients extracted from the soil by the roots whilst the latter need the products of photosynthesis from the shoots in order to grow further. Root growth is often under-estimated, for instance it is an astonishing fact that the total root length, including root hairs, of COUCH GRASS (*Agropyron repens*), with its fibrous root system, has been calculated at over 500 kilometres (300 miles)!

All living bodies need an internal transportation system to circulate nutrients and waste products to and from the vital organs, and just as animals have a network of veins and arteries, so too do plants. The plumbing system of a plant is carried in its stem, a structure which must also support the weight of the aerial parts and arrange them to gain maximum benefit of the light available. In the stem are the veins, or more properly, the *xylem elements*, which form a continuous pipeline from the root tips to the top of the plant, up which the nutrients in solution are brought. Food materials manufactured in the leaves by photosynthesis are transported about the plant to wherever they are required, for growth or storage, by the plant's 'arteries', called *phloem sieve tubes*. The efficiency of the plant's plumbing system is such that no single cell anywhere in the leaf is more than ten to twenty cells removed from some part of the system to supply it with essential water. Movement of water in the xylem is achieved by root pressure – that is, it is forced up by pressure from the water-absorbing root cells, – together with the cohesive force of water (the force which causes molecules of water to cling tightly together, and resist being pulled apart), which draws water up through the xylem from the roots as water is lost through transpiration at the top of the plant. This force is so strong in a plant cell that it enables water to be drawn up to the top of the tallest trees. For plants, though, life proceeds at a gentle pace; water flows in the stem at a rate of 20–30 centimetres (8–12 inches) per hour, a distance which can be covered in less than a second by blood in the veins of an animal.

A germinating seedling of a higher plant, then, given its basic requirements of water, air, light, warmth and nutrient availability, in whatever proportions are necessary to the well-being of the particular species, is likely to grow roots, shoots and leaves, by a process of cell division, gradually building up into the typical shape of the plant species. The rate of growth is governed by special plant growth hormones, substances so potent that 30 grams (a single ounce) could produce enough growth, laid end to end, to circle the earth. Natural plant growth rates vary; [species of bamboo such as *Bambusa arundinacea* can grow as much as 90 centimetres (3 feet) in a day, though a rate of 1–5 centimetres (½–2 inches) is a more normal maximum rate for most plants.] Cell division occurs in special regions at the root and shoot tips; woody plants can increase their stem girth too by special methods of annual thickening, to produce the ancient massive trunks of, for example, our old oak trees.

Its growth pattern completed, and conditions being favourable, the mature plant is ready to unveil its ultimate achievement – the flower.

2: The functions of flowers

THE FLOWER, THAT MOST CHARACTERISTIC OF ALL angiosperm features, heralds the start of the sexual reproductive process of the plant, a process so successful that it is largely responsible for our great variety and widespread distribution of wild flowers. The function of the flower in the life of a plant is the production of seed, and its colour, scent, elaborate parts and intricate mechanisms all help to achieve this end.

Beautifully varied though they may be in colour and form, all flowers still share a common plan. At the base, supporting the showier parts above, lies the *receptacle*, the expanded end of the flower stalk. The green leaf-like *sepals* come next, protecting the flower bud and later curling back as the flower opens to reveal the splendour of its petals. The flower parts are, in fact, leaves that have become modified for their particular functions. The petals act to advertise the presence of the flower to insects with their often showy arrangements of colour and form, and frequently provide scents and sweet nectar too. Sometimes sepals can also became coloured and help in the advertisement display, a situation often found amongst the flowering monocotyledons, such as the orchids. Sepals and petals together surround the sexual part of the flower, the function of which is to produce, bear and protect the sex cells. The stalked *stamens* comprise the male part. They make pollen grains within their usually bright yellow swollen ends, the *anthers*. The pollen is held in the four little sacs of the anther until it is ripe, when the sacs rupture. Right in the centre of the flower lies the female part, the *ovary* containing the ovule, later to become the seed after fertilization. On the top of the ovary, and connected to it by a little stalk called the *style*, sits the sticky *stigma*, the receptive surface.

For a seed to develop, male and female cells, or at least their nuclei, must meet and fuse in the process known as *fertilization*. In order for this to happen, male pollen must first reach a ripe female stigma of the same species; this is known as pollination. Self-pollination results from pollen passing from an anther of a flower to the stigma of the same flower or another flower on the same plant, while transportation of pollen to the stigma of a flower on another plant (of the same species) is known as cross-pollination. Plants have devised many ingenious and elaborate mechanisms to ensure or at least favour the latter case. Cross-pollination makes sense since, by having two different parents, the chances of variation amongst the offspring are much increased, thus setting the stage for natural selection to act. A different or changing environment would favour those individuals better adapted to fit that

Typical longitudinal section through a flower

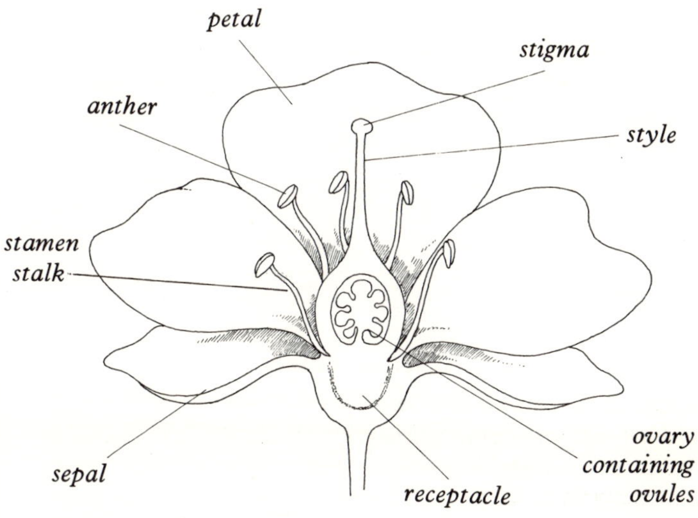

environment, whilst the others would tend not to survive in the course of competition for their primary needs.

Ovules are fixed inside their ovaries. Pollen grains cannot move either, so outside agencies have to be employed to carry them to their destination. In temperate climates, insects and the wind are the most important pollen-bearers, though water and slugs and snails can sometimes be useful. Bat-pollination is quite important in the tropics, where the flowers designed for this method are large and strong to guard against damage by bat claws, with dingy colours and sour, musty scents similar to the bat's own glandular secretions. Such a plant is the Indonesian FREYCINETIA INSIGNIS which provides fleshy sugary leaves just beneath the flower for its fruit-bat pollinators to eat. Bird-pollination too is common in warmer countries, hummingbirds being especially important. Such flowers have no scent (birds have a poorly developed sense of smell), are often garishly coloured as in the green, scarlet and yellow 'parrot' colouration often found in the pineapple family. They provide a lot of nectar to satisfy their bird visitors.

Insects, however, are by far the most useful pollinators. Their small size and hairiness makes them trouble-free and efficient visitors. Flowers adapted for this method lure the insects with their brightly coloured, often scented, petals; and, if the flowers are small, they are often massed together on the plant to increase the visual advertisement display. They also provide sweet nectar to make the visit worthwhile for the insect, and have pollen which is spiny and sticky so that it will cling to the insect's hairy body. Many have special grooves or darker lines (honey-guides) pointing towards the centre of the flower directing the insect inwards to find the nectar, and incidentally the sex organs. Sometimes these honey-guides are only visible to humans under ultraviolet light, for bees at least are able to see ultraviolet. After-dark pollinators like moths are attracted by large pale-coloured petals, more visible in the dim light than other colours. Night-opening flowers are often strongly scented too, like the large yellow-flowered EVENING PRIMROSE (*Oenothera odorata*). Other flowers such as the WILD ARUM (*Arum maculatum*), or 'cuckoo pint' are designed to attract flies as pollinators and smell quite repulsive to our noses. Indeed, certain African plants are called 'carrion flowers' because they smell, as well as look, like bad meat. Interestingly, insect ancestors living 250 million years ago, a time when the flowering plants did not exist, had little sense of smell; they seem to have developed it in close association with the evolution of the flowering plants.

These flowers contrast strikingly with those of plants such as the oak or hazel which have their tiny, individually inconspicuous, male flowers crowded together into dangling yellow catkins packed with pollen, to be released like clouds of gold-dust on the wind. For these wind-pollinated flowers have no need of attractive petals, indeed they would be an encumbrance, impeding the wind in its purpose. Scents and nectar are lacking too, and the flowers generally appear well before the leaves or are held well above them, so that there is

An insect visiting a flower for nectar

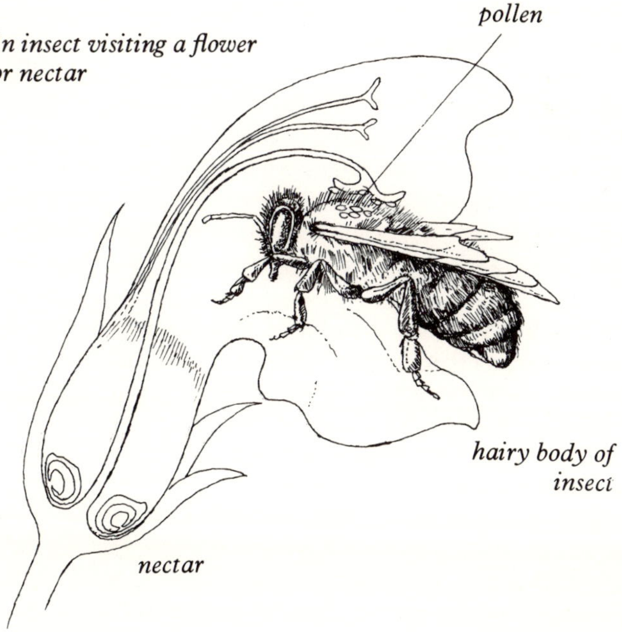

A grass flower, showing its adaptations to wind pollination

no interference with the free circulation of air around them and the release and capture of pollen grains. They need to produce a lot of pollen because the wind is rather a risky and erratic means of transport as far as pollination is concerned, and their pollen grains are usually very tiny, smooth-walled, dry and light so that they can remain air-borne as long as possible. Sometimes they have little air sacs too. The female stigmas, often large and especially sticky, reach well out of the flower to increase the chances of catching the air-borne pollen.

In our few water-pollinated species, flower petals are again reduced as in the EEL-GRASSES (*Zostera* species) which has thread-like pollen of the same density as sea water, enabling them to move freely at any depth until caught upon the feathery stigmas of the female flowers.

Some plants, such as WOOD SORREL (*Oxalis acetosella*), have flowers which never fully open and so pollinate themselves, the so-called *cleistogamous* flowers. Others, like the sweet violet, have ordinary 'open' flowers adapted to insect pollination and little cleistogamous flowers as well, as the former flowers set seed poorly. Yet others can resort to self-pollination should cross-pollination fail as in LOVE-IN-A-MIST (*Nigella*) which initially has the stamens held up high out of the reach of its own pollen but later, in the absence of cross-pollination, the styles bend over backwards until the stigmas are brought into contact with the pollen.

However, most plants favour cross-pollination as we noted earlier, and numerous mechanisms exist in nature to further this possibility. Apart from having the stamens and stigma so positioned that self-pollination is unlikely, many plants like the daisy and DANDELION (*Taraxacum*) have male and female parts which ripen at different times, whilst other plants completely separate the sexes, having either one-sexed flowers on the same plant or on separate ones. This situation is common amongst wind-pollinated flowers where the vast amounts of pollen produced would all too easily smother the stigmas if they were close by. Often though self-pollination is prevented by sheer incompatibility or self-sterility, as in the buttercups.

Whatever the means, successful pollination results in the pollen grain pushing out a little tube which, having a natural tendency to grow away from light and air, pushes down through the female tissues until it reaches the ovule in its ovary. In corn (maize), the pollen tube must grow down the length of the silk and the whole length of the ear to reach the ovule, some 30 centimetres (1 foot) in fact, and this from a pollen grain measuring little more than one-four-hundredth of a centimetre ($\frac{1}{1000}$th of an inch)! Upon penetrating the ovule, the tube tip discharges two male nuclei, one of which fuses with the female ovule nucleus to form the fertilized ovum, the other fusing with two special ovule nuclei and thus initiating the *endosperm*, the tissue which will surround and nourish the seed embryo.

The seed ripens, becoming enclosed in a tough coat, for it will shortly be employed as a portable strongbox for the delicate embryo being nurtured inside. A tiny root (*radicle*), shoot (*plumule*) and embryonic seedling leaves (*cotyledons*) form, and the whole seed undergoes a drying-out process so that its water content is reduced to around 10% by weight. As its seeds develop, the flower as we know it withers and falls away, leaving the protective ovary to enlarge and form the fruit, which may be soft and fleshy like sloes and blackberries, or hard and dry like acorns and broom pods. Sometimes the fruit can develop from other floral parts too, such as the receptacle as in the case of the rose hip.

The reproductive process that began with the formation of the flower has reached its final stage: the dispersal of the ripe seed from the 'mother' plant to a site suitable for germination. Towards this end, the ripe fruit and seeds have become specially adapted. Some fruits split and eject their seeds themselves, many others harness again the aid of the pollination agencies of passing animals, wind and water; and, in fact, there are almost as many fascinating devices associated with seed dispersal as there are with pollination. They range from the explosive pods of the broom, to the lazily drifting parachutes of the dandelion family, from the attractive berries regurgitated by over-indulgent birds (many seeds incidentally, survive the journey through

How a seed is made

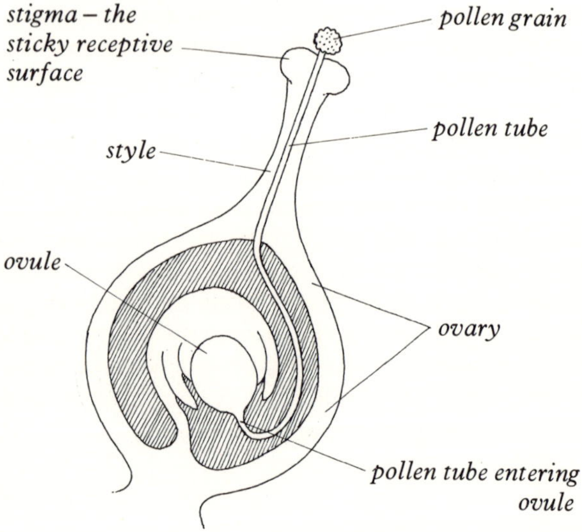

In goose grass, or 'cleavers' (*Galium aparine*) the ovary wall develops hooks. These may catch in the fur or wool of an animal's body.

The fruit of the ash (Fraxinus excelsior)

has a long, twisted wing which will send the enclosed seed spinning on its way.

an animal's gut; some are even in a better state for germination afterwards!), to the membranous winged fruits of ash and sycamore sailing gently groundwards, and from the capsules of the common poppy which shoot their seeds out like pepper from a pepper pot as the wind flips their stalks, to the familiar burrs of GOOSE-GRASS (*Galium*) designed to cling to the fur of passing animals, or woolly stockings of people! Some plants are adapted for dispersal by ants and others even, apparently, by man's own vehicles of transport!

The flower, then, sets in motion a series of intricate and marvellous activities with but one ultimate aim, the continuation of the species.

3: The ecology of plants

PLANTS CAN BE BETTER UNDERSTOOD IF WE CONSIDER them in relation to their environment; this study is called *ecology*, a word derived from two Greek words meaning 'the study of the home'. We saw earlier that plants do not live in isolation but form communities, the structure of which varies with the environment in which the members of the community live. Thus we find completely different plant species growing together in a wood to those beside a stream, or on a roadside verge. Within all except the simplest communities, the species can be found arranged in layers, or strata, of appropriate life-forms, as in a wood with its four distinct layers formed by its trees, shrubs, herbs and mosses.

The study of a plant's home would therefore include its climate, soil and also the other associated organisms, plants and animals (including man). Many inter-relations exist here, however. For instance the latter (so-called *biotic*) factor would include conditions such as seed availability and viability, factors which, as we have seen, are intimately connected with methods of plant reproduction, which have evolved in association with the availability, or otherwise, of particular pollination agencies, be they climatic or biotic. Each plant species thrives best within certain limits for each of these environmental factors, the complex meshwork of which serves to ensure that only plants adapted to survive the prevailing conditions survive to become components of the community.

The major controlling influence on vegetation is the

The layers of a wood from top to bottom: Tree layer – oak in very young leaf; Shrub layer – hazel; Herb or 'field' layer – wood anemone, primrose, sweet violet, dog's mercury; Moss layer.

climate, a general term covering such important factors as temperature, light humidity and wind, though of course these do not act independently of each other. Climate can easily be seen to be primarily responsible for such recognizable plant formations or biomes of the world as the coniferous forest of colder regions with its evergreens especially adapted to utilize the available light, water and mineral nutrients peculiar to these colder regions, or the tropical rain forest with its varied rampant evergreen growth forms promoted by the hot, humid climate, with its all-year-round growing season. In addition, though, the character of an area of vegetation can change quite dramatically as a result of variations in the lay of the land, its topography, giving rise to localized climate variations, within a larger area of prevailing constant climate and soil conditions. The southern slopes of a mountain-side, for instance, will harbour mainly drought-tolerant and warmth-loving species, whilst plants requiring cooler and moister conditions would be found on the northern inclines. Again, temperature inversions may exist in deep gorges, enabling characteristically alpine species to grow at low levels with warmth-loving species above them.

Soil types can vary greatly in their physical and chemical compositions as determined by the basic 'parent' rock, from which the soil originally weathered. Especially important is the amount of accumulated organic matter (*humus*) the soil contains, and its associated water-holding capacity. These factors, associated with climatic conditions, control another basic characteristic of the soil, its acidity or alkalinity. The importance of the soil factor in the environment can be seen in the comparison of a well-drained soil with a heavy rainfall supporting a rich forest, with a less-porous soil and the same high rainfall and its subsequent development into a peat bog.

Part of the essential nature of a plant community is its instability in that, under perfect conditions, the plant cover will develop from an original pioneer community, through a succession of recognizable stages to a 'climax community'. Such a succession can occur from bare rock inhabited by bacteria, blue-green and green algae, which break down the surface rock and deposit the first dead organic matter, which with the aid of the elements, comes to form a 'soil', allowing secondary colonization by lichens and mosses, and later of higher plant forms, including trees. A climax community is basically the highest form of plant life which may be supported by a particular environment; it is said to be in equilibrium with those environmental conditions at that time. Thus, deciduous woodland is the natural climatic climax in England. Some vegetational successions, however, appear to have reached a temporary stability where arresting factors are involved; these may be to do with the soil as in bogs where tree-growth is prevented by the soil being too wet, or, again, animal (often human) influences may be at work.

4: Flowers of the coast

THE COASTLINE OF OUR ISLANDS HOLDS BEAUTY AND variety at every turn – shingle beaches and imposing spray-blown cliff-faces, near-flat stretches of salt marsh and the wind-built undulations of marram-covered dune systems, often lying close to one another, and each bearing its own characteristic flora, including both the fantastic and the beautiful. Coastal vegetation has one major linking factor, however, and that is its tendency to be salt-loving, or *halophytic*, and succulent bodies are a common feature here, with their large cells distended with salt solutions. Coastal habitats are of their very nature ever-changing, the sea gradually erodes the cliffs and piles mud and sand up onto the salt marshes, and winds pile up sand on the dunes, and coastal plant communities, in adapting to these changes, occupy marked zones, with the succession of species very much apparent.

Sand-dunes are especially interesting in this respect. Often built upon shingle, they can reach 12–13 metres (40–60 feet) in height, and their mobility is such that whole villages can be buried by them, given a century or so. Their early development begins just landward of that arid strip of foreshore which lies beyond the reach of ordinary high tides, with its 'drift-line' belt of succulent annuals like the pretty lilac-flowered SEA ROCKET (*Cakile maritima*). SALTWORT (*Salsola kali*), with its needle-sharp spines and history of alkali production for glass and soap-making, is found scattered amongst the dead seaweed which provides humus for the spray-drenched soil. Sometimes, too, three close relations of our familiar vegetables can be found here: SEAKALE (*Crambe maritima*), SEA BEET (*Beta maritima*) and SEA RADISH (*Raphanus maritimus*).

Dunes form by the accumulation of wind-blown sand around objects and plants, which effectively stop the wind in its path, causing it to drop its load of heavier particles, most particularly on the lee side. The embryo dune thus formed can further increase in size if the object is a plant which can push its shoots upwards and sideways as it becomes buried, so that the obstacle in effect grows 'on demand'. The drift-line species mentioned above can build little 'dunelets' but, as they are unable to push through the sand, there is no increase in size here. It is only where the important perennials, SEA SANDWORT (*Honkenya peploides*) and SEA COUCH GRASS (*Agropyron junceiforme*) can gain footholds in the shifting sands that the fore-dunes proper can begin to form.

Both species are able to grow through when buried, though *Honkenya*, with its fleshy leaves which used to be eaten as a pickle, forms rather small and impermanent dunes. *Agropyron*, too, has a limit to its upwards growth for the typically flat-topped dunes it forms are rarely more than 1.5 metres (4 feet 7 inches) high, but its extensive roots and creeping underground stems push deeply down and create stability; it is a successful pioneer dune-former in fact, and one that is tolerant of immersion in salt water, unlike another, most important, dune-forming grass – marram grass.

Marram, with its horizontal shoots bearing whip-like leaves tightly rolled in dry weather, has roots which can penetrate to a depth of 18 metres (60 feet), and thrives best in the main dune ranges where the sand is still fresh and mobile. Only marram can continue growing upwards for many metres through an ever-increasing mound of sand, constantly forming fresh roots in the moist sand just below the surface, to form massive, typically dome-shaped dunes of great height. In between the marram tufts, the distinctive bluish flowers and grey spiny leaves of the SEA HOLLY (*Eryngium maritimum*) can often be found. This too can recover from burial and has an extensive root system of up to

Sea holly (Eryngium maritimum): *this striking sand-dune plant, with its spiny flower-heads and bluish-grey, thistle-like leaves, is often grown in gardens for decorative purposes. It is well adapted to the dry sands of its true home for its leaves are so tough and leathery, that water loss through them is reduced to an absolute minimum.*

2 metres (6 feet 8 inches). Other plants begin to appear; the CREEPING and SPEAR THISTLES (*Cirsium arvense* and *C. vulgare*), HOUND'S-TONGUE (*Cynoglossum officinale*) with its curious tiny brownish-red flowers, and another dune-forming grass, LYME-GRASS (*Elymus arenarius*), which dislikes high winds. Once formed, marram dunes are invaded by several species capable of matting together the surface particles with their creeping rhizomes, like the CREEPING FESCUE GRASS (*Festuca rubra*), *the* SAND SEDGE (*Carex arenaria*) and the COMMON SEA SPURGE (*Euphorbia paralias*). The SEA CONVOLVULUS (*Calystegia soldanella*), with its fleshy leaves and beautiful flowers like little pink trumpets, helps to stabilize the sand a little too.

Once fixed, the dunes show a change in their vegetational cover again. Marram becomes less frequent as it needs fresh sand to keep it healthy, and carpets of certain mosses and particularly lichens, enter into the picture, lending an overall greyness to these fixed dunes, in contrast to the younger 'yellow' dunes. The soil, too, of older dunes, itself looks grey as it contains more humus from the decayed remains of earlier colonizers and can now support an almost continuous plant cover, with many species of flowering plants, especially heath and dry-grassland species, like the COMMON STORK'S BILL (*Erodium cicutarium*), YELLOW STONECROP (*Sedum acre*) and SHEEP'S SORREL (*Rumex acetosella*). Health gives way to shrubs, including BLACKBERRIES, ELDER, HAWTHORN and WILD ROSES.

Dunes are characterized by their instability, lack of

Sea convolvulus (Calystegia soldanella)*: this beautiful flower may be found trailing its pink or pale purple funnel-shaped blooms over many southern sandy shores. It is pollinated mainly by humble-bees, but is sometimes self-pollinated.*

Yellow horned poppy (Glaucium flavum)*: the handsome golden-yellow flowers of this plant may be up to 8 centimetres (4 inches) across. The leaves, like those of the sea holly, are tough and leathery; they, and the branching stems, have a characteristic bluish-green colour ('glaucous'). The seed-pods are very strange; very long, and stalk-like, they are the 'horns' of the plant's English name.*

nutrients and extreme surface dryness, and plants adapt either by 'digging their heels' in with deep roots, or by the 'disappearing tactics' of minute annuals like the MOUSE-EAR CHICKWEEDS (*Cerastium* spp.), the early HAIR GRASS (*Aira praecox*) and SAND CAT'S-TAIL GRASS (*Phleum arenarium*), which are adapted to live only during the time of the year when the dune surface is moist for they germinate in autumn, flower and fruit by Whitsuntide, die, and the seeds lie dormant until autumn.

Shingle vegetation, usually sparse and localized as typified by Chesil Bank in Dorset, differs from that on dunes as mostly the plants are quite widespread species like the SOW THISTLES (*Sonchus arvensis*) and (*S. oleraceus*), RAGWORT (*Senecio jacobaea*) and GROUNDSEL (*Senecio vulgaris*), all common inland weeds. An interesting shingle plant is the SHRUBBY SEABLITE (*Suaeda fruticosa*), a small shrub with many branches which not only withstands being covered by shingle, but appears to be stimulated by it too! Other colonizers of this habitat include the YELLOW HORNED POPPY (*Glaucium flavum*) with its fancy yellow flowers followed by long seed-pods, and the SEA CAMPION (*Silene maritima*), which has pure white flowers in June. Sea sandwort is commonly found here too. In this exposed environment, winter winds pose a problem which shingle plants solve in various ways; annuals like groundsel, overwinter as a seed, the deep-rooted poppy exposes only a ground-hugging rosette of leaves and the spreading nature of the sea campion causes it to rise little above the shingle surface, whilst the sea sandwort disappears completely below the ground. Shingle plants do not suffer from drought as dew forms overnight between the pebbles. Terneries are quite common on shingle and these cause localized patches of rich plant growth as a result of the boost in nutrients derived from the birds' waste products.

On flat stretches behind spits of protective shingle, where the wave action is reduced, mud and silt brought in on the tides often collects and forms salty marshes, a peculiar environment for plants as they are frequently covered in salt water, twice a day in fact on the lower levels, so the vegetation must be truly halophytic.

There are very definite and characteristic areas of vegetation in salt marshes which reflects the variation in frequency of submersion. The lowest levels frequently harbour the grass-wracks or eel grass, a sloppy-looking monocotyledon which flowers under water and has water-borne pollen. Its flexible leaves, when dried, are excellent for stuffing mattresses! A little higher, the GLASSWORTS, or 'marsh samphires' (*Salicornia* spp.), with their fleshy, segmented, leafless stems, gain a footing and often proceed to cover large areas. Their roots begin the stabilization of the mud, allowing other plants to join them. Nowadays, the CORD-GRASS (*Spartina townsendii*) often appears at this stage. Though only recently introduced into Britain, its fast growth and efficiency in soil stabilization (its 'hybrid vigour', in fact), has ensured its successful colonization of vast areas of salt marshes around our coasts. This, naturally, has upset the balance of life existing previously on the marsh, and ducks like the widgeon, which feed on eel grasses, often go hungry.

Zonation on a salt-marsh. Left to right: eel grass (Zostera)*; marsh samphire* (Salicornia)*; sea aster* (Aster tripolium)*; sea lavender* (Limonium)*; sea purslane* (Halimione portulacoides)*; mixed marsh turf, comprising thrift* (Armeria maritima)*; sea plantain* (Plantago)*; rushes* (Juncus) *etc.*

SEA MANNA-GRASS (*Puccinellia maritima*) forms the next band, though some marshes have instead a zone of SEA ASTER (*Aster tripolium*), its heads of flowers bright with their lilac rays and yellow centres, though there is a form (var. *discoideus*) which lacks rays. The purple theme is taken up and spread over the higher marshes as the SEA LAVENDERS (*Limonium* spp.) and then the SEA PINKS, or 'thrift' (*Armeria maritima*), come into their own, along with SEA PLANTAIN (*Plantago maritima*), RUSHES (*Juncus*) and many others forming a mixed community marsh turf upon which sheep and cattle often graze, and this blends in with the non-halophytic vegetation of the neighbourhood. Sometimes the drier upper levels are carpeted closely to a depth of half a metre (1 foot 8 inches) or so with the silvery-grey foliage of a very characteristic marsh plant, the shrubby SEA PURSLANE (*Halimione portulacoides*), frequently found, too, lining the banks of deeply cut drainage channels.

Sea plantain, lavenders and thrift are commonly found on cliff faces too, luxuriating here in the salt spray. Wind- and salt-tolerant species from the hinterland intermingle here with the true coastal plants, like the notable cliff-lover, or *chasmophyte*, ROCK SAMPHIRE (*Crithmum maritimum*), with its clinging cushions of divided leaves which used to be favourites for pickling. The SEA CARROT (*Daucus gingidium*) and SEA CABBAGE (*Brassica oleracea*), with its bright yellow flowers set against a backdrop of grey Portland limestone, present a fine sight on the south Dorset coast and are among the many interesting plants to be found along our cliffs.

5: Flowers of woods and forests

THOUGH OUR WOODLANDS AND FORESTS ARE BUT 'A patch' compared to the enormous areas they occupied in earlier days, Britain is fortunate in possessing fine woodlands harbouring a variety of species. Of these, the PEDUNCULATE OAK (*Quercus robur*) with its stalked acorns, with HAZEL (*Corylus*) as the dominant species in the shrub layer, is truly the heart of British woodland; it is our climatic climax vegetation, and its herbaceous or 'field' layer provides the finest show of spring flowers, before the oaks come later into leaf.

Perhaps the best known and most abundant of these is the delicately scented primrose, a plant which favours good soils, as a distribution study in Dorset clearly showed. The plant was found to be virtually absent within a band running north-east across the centre of the county, a band which was later correlated with the distribution of soil types and found to follow the line of course of shallow chalky or sandy soils. Oak woods in the south-east of England, where there is too much water and lime in the soil for primroses, have instead carpets of OXLIPS (*Primula elatior*) with their primrose-like flowers clustered on a tall stalk.

Keeping company with the primrose on fertile woodland soils are sweet violets and wood anemone starring the floor with its elegant white flowers, and the less attractive but very prolific DOG'S MERCURY (*Mercurialis perennis*) whose shoots begin to grow before the winter is over with leafy stems persisting into and through the summer. Here, too, SANICLE (*Sanicula europaea*) grows in abundance with its shiny leaves and little starry pinkish-white flowers lighting up the dense shade. It was a herb much used for wound-healing, hence the old saying 'He who keeps sanicle laughs at the doctor'. Here and there, certain other species can become locally dominant, including the WILD STRAWBERRY (*Fragaria vesca*), the pinkish-white-flowered ENCHANTER'S NIGHTSHADE (*Circaea lutetiana*) and the little ground ivy, sometimes called 'Gill-go-over-the-ground' and used by country folk to make 'Gill tea' for warding off colds

Primrose (Primula vulgaris): *'Prima rosa' means 'the first rose'; the primrose is certainly one of the earliest spring flowers. The flowers show an interesting adaptation to cross-pollination; they are of two distinct kinds, never found together on the same plant. Some primroses have short stamens with the green pin-like knob of the stigma towering above them and visible at the top of the flower-tube ('pin-eyed'), whilst others show the stamens at the top of the tube, and have the stigma lower down ('thrum-eyed'). A long-tongued insect sucking nectar from the tube-bottom of a pin-eyed flower will pick up pollen on its mouthparts at a level which will be at the right height for deposition onto the stigma of a thrum-eyed flower alone.*

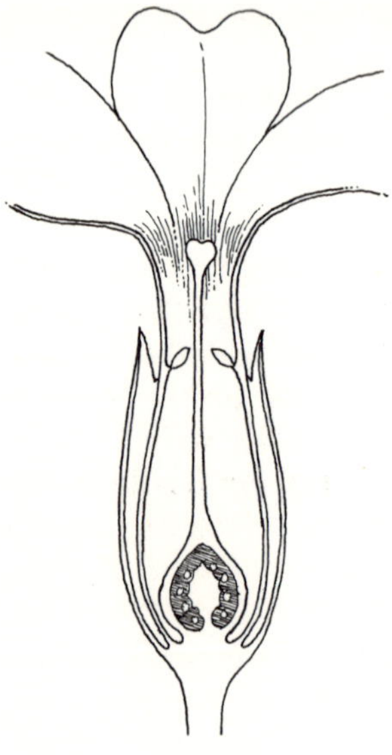

pin-eyed

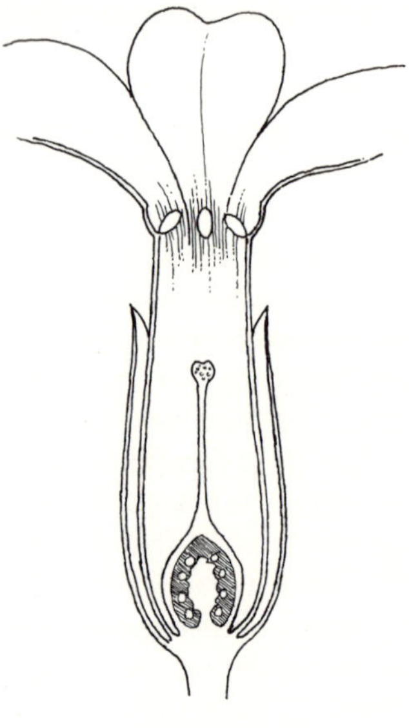

thrum-eyed

Sweet violet (Viola odorata)*: there are many species of violet. Some are pale-flowered, with hairy leaves, many are scentless, but the finest of all is the sweet violet. This little flower has been used for centuries in the making of perfumes and love philtres; as a dye, and candied as a favourite decoration for sweet foods. The fresh leaves are sometimes made into a compress and placed on sore throats to soothe them.*

and coughs. Scattered plants of many other species are often found growing in these fertile oak woods, including BUGLE (*Ajuga reptans*), HERB-ROBERT (*Geranium robertianum*), wood sorrel (sometimes called 'cuckoo's meat', with its pale pink flowers), ARCHANGEL or 'yellow dead-nettle' (*Galeobdolon luteum*) and BURDOCK (*Arctium minus*). The abundance of the latter in places is probably due in no small way to its tenaciously clinging burrs and its distastefulness to animals.

Most of today's woodlands and forests are the result of tree-felling and replanting by man; another of his activities, periodic coppicing of the hazel in oak woods, profoundly affects the field layer in particular, since the dense shade preceding coppicing excludes all but a few flowers, whereas a few years later many species have re-invaded the less-shaded terrain. Some species can 'hang on' in deep shade, and then make a great show of flowering when they find light again, as with the lovely rosebay willowherb, with its creeping underground rhizomes, and the FOXGLOVE (*Digitalis purpurea*), both conspicuous flowers of woodland clearings. Rosebay, with its masses of silky plumed seeds, may be borne for great distances and often colonizes fire-cleared areas, hence its nickname of fireweed.

Open oak wood on drier, lighter soils has a field layer dominated by BRAMBLE (*Rubus fruiticosus*) with bracken crowding-out all but a very few species. One of these is the sweetly scented HONEYSUCKLE (*Lonicera periclymenum*) which, with its 3–6 metre (10–20 feet)-long stem, can twine its way into the light. These soils are favoured too by the bluebell, whose shoots appear in February or March, before those of the SOFT-GRASS (*Holcus mollis*) with which it is often associated, and well before the bracken, and as their roots occupy

The plumed seeds of rosebay willowherb (Epilobium angustifolium) *are readily air-borne and can travel considerable distances in this way.*

different layers of soil, serious competition is avoided. Another common spring flower is the lesser celandine with its eight-pointed, star-like golden flowers, a favourite of Wordsworth who wrote of its 'glittering countenance' and has the plant engraved on his tomb. The specific name 'ficaria' derives from its little root tubers, each capable of producing a new plant, which resemble figs.

Mixed ash-oak woods are often found on chalky soils, and here the dainty COLUMBINE (*Aquilegia vulgaris*), with its drooping purple-blue flower-heads, and the fine NETTLE-LEAVED BELLFLOWER (*Campanula trachelium*), the original Canterbury bell, can be found. Locally common, too, is the MEADOW SAFFRON (*Colchicum autumnale*), its pale flowers appearing in late summer, well after the leaves have faded, hence its local name 'naked ladies'. The corms and seeds of this plant contain colchinine, a drug employed from early times in pain-relief. STINKING IRIS (*Iris foetidissima*), with its big red seeds and leaves that smell of garlic when crushed, can also be found in chalky soiled woodlands.

Yew and pine woods cast such dense shade that little can grow beneath them. A young beech wood too, with its deep leaf-canopy, gives the impression of an ill-lit, thinly congregated cathedral, but its deep humus harbours some fascinating saprophytes (plants lacking in chlorophyll and feeding directly off dead organic matter, aided by fungi), like the BIRD'S-NEST-ORCHID (*Neottia nidus-avis*), which, with its yellowish-brown colouration, looks almost dead, and the paler YELLOW BIRD'S-NEST ORCHID (*Monotropa hypopitys*). Two other orchid genera, the HELLEBORINES (*Epipactis* and *Cephalanthera*), are also characteristic here. The field layer of older, more open-canopied beech woods, has a richer flora, similar in species composition to the oak woods on fertile soils.

Woodlands to be found on waterlogged soils contain mostly marsh species, plants considered in the following chapter.

6: Flowers of marshes and fens

THE FLORA OF MARSHES AND FENS IS VERY SIMILAR. BOTH habitats share the conditions of a waterlogged, neutral to alkaline, soil which is usually rich in mineral nutrients. The difference between the two can be seen largely in terms of peat, which is the accumulated mass of undecomposed plant remains that develops in waterlogged soils; in marshes this development is usually prevented either by erosion or conditions of free drainage, whilst fens, in their low-lying basins, form layer upon layer of peat in their drainage water which usually comes from surrounding *calcareous* (calcium bearing, chalky) rocks, as in East Anglia with its chalky uplands. A third distinctive type of flora arises under conditions of peat accumulation on waterlogged *acid* soils, that is, bogs, habitats characteristically found in non-calcareous uplands with a cool and rainy climate, as in the north-west of Britain.

Marsh plants, the most characteristic of which is the rush, are often found along the edges of rivers or around ponds or lakes, behind the reedswamp zone, which is usually dominated by the COMMON REED (*Phragmites*). It is frequently a much-disturbed area, being grazed and trampled by cattle, and where the grasses and herbs of neighbouring drier pastures intermingle, cut for hay. Common species of these 'water meadows' include that troublesome garden weed, CREEPING BUTTERCUP (*Ranunculus repens*), the coarse TUFTED HAIR-GRASS (*Deschampsia cespitosa*), the familiar RAGGED ROBIN (*Lychnis flos-cuculi*) amongst tall drooping sedges. 'Queen of the meadows', otherwise known as MEADOWSWEET (*Fili-*

pendula ulmaria), a plant held sacred by the Druids, is a frequent member of the community. With its finely aromatic leaves and flowers, it is a plant still used today in the making of herb beer.

In the natural course of events, a marsh would be colonized by woody plants like the sallow and then trees, mainly willows, birch, ash and alder; alderwood being the climax vegetation of marshes. In olden days, alders were encouraged by planting and were then coppiced to be used in the manufacture of old black gunpowder and, too, in northern England, for clogs!

Ragged robin (Lychnis flos-cuculi)*: the rosy, toothed petals of this frequenter of moist places protrude from a reddish, purple-veined tube formed by the sepals. The flowers have no scent, but provide lots of nectar for which they are visited by butterflies and long-tongued bees and flies. The stamens produce their ripe pollen well before the stigmas are mature, so that cross-pollination is favoured. Occasional white forms of this flower may be found. Its name is derived from a Greek word 'lichnos' meaning a light or lantern, because it is such a bright flower.*

Yellow flag (Iris pseudacorus)*: this stately British flower, commonly found fringing rivers and lakes, is the original of the famous French royal emblem, the Fleur-de-lys. The showier parts are again formed by the sepals. They are borne on tall stalks arising from a thick creeping rootstock which contains a bitter sap supposedly useful in curing coughs, toothache and snake bites. The pale-green leaves are so sharp-edged that they can quite easily cut your hands.*

Marshy woodlands often contain BLACK CURRANT (*Ribes nigrum*), RED CURRANT (*R. rubrum*) and GOOSEBERRY (*R. uva-crispa*), mixed with IVY (*Hedera helix*) and a number of other climbers, a very characteristic plant form here, lending to the habitat the appearance of a miniature-scale damp, tropical forest. The commonest climbers are the GREAT BINDWEED (*Calystegia sepium*) with its big white trumpet flowers, HOP (*Humulus lupulus*), and from the family of the potato and tomato, BITTERSWEET or 'woody nightshade' (*Solanum dulcamara*) which after the winter's rest can grow to a height of 2–2.5 metres (6 feet 8 inches–8 feet) to bear its glossy red berries, commonly cut open and bandaged upon sore places by country-folk at one time. Ground scramblers flourish too, such as the COMMON CLEAVERS (*Galium aparine*), so brittle-stemmed that shoots as well as its clinging fruits snap off and stick to passers-by. Its seeds are occasionally roasted and made into a coffee, which is apparently mildly stimulating! Other characteristic shade-tolerant marsh plants of the field layer include MONEYWORT or 'creeping jenny' (*Lysimachia nummularia*) with its round flowers like golden coins, various sedges, YELLOW FLAG (*Iris pseudacorus*), STINGING NETTLE (*Urtica dioica*), HEMP AGRIMONY (*Eupatorium cannabinum*) and COMFREY (*Symphytum officinale*) or 'knit-bone', used in herbal medicine for inflamed areas, and once a popular fodder crop too.

Fen communities are a stage in the natural succession from water to land vegetation; they develop on peat derived from the accumulated annual remains of the stems of reed-swamp plants, as opposed to the mineral soils of marshes, and are therefore found typically in flat lakeside and riverside areas where there is very little silting. The vegetation, however, is as we noted earlier, very close to marsh vegetation. Once the peat surface is free from standing water, shrubs and trees tend to invade forming what is known as *carr* (connected with the Icelandic word 'kjarr', meaning a fen wood). If, in its early stages, this carr is mown, development is arrested and the herbaceous fen community is thus perpetuated. Thus the 'natural' fen vegetation of, for instance, Wicken Fen in Cambridgeshire, can only be kept as such by human activities; the reeds which dominate most fen communities are periodically cut for thatching, whilst the rest, mainly species of sedge and grass, are used for litter, which is the fenman's cattle bedding.

Sedges are classic fen plants. Though their flowers are small and inconspicuous, being adapted to wind-pollination, they are actually quite an interesting family. They may be distinguished from grasses by their triangular stem cross-section and where leaves are developed, these are arranged in three planes, seen from above, as opposed to two rows and a round stem in

Woody nightshade (Solanum dulcamara)*: also known as bittersweet, the poisonous woody nightshade drapes itself around other plants in damp places and hedgerows. Its leaves vary considerably ; the lowest are heart-shaped, the upper ones spear-shaped. Like the potato, its flowers may be either lilac or* (*rarely*) *white. The egg-shaped berries, often found with flowers on the same plant, are green at first and gradually change through shades of yellow and orange until the fine, jewel-bright red of the mature fruit is reached in early autumn.*

A similar plant, the extremely poisonous deadly nightshade (Atropa belladonna) *has big, flat, black berries and larger bell-shaped flowers. It is much less common than bittersweet.*

grasses. The genus *Carex* contains members ranging in size from a few centimetres (an inch) to giants of 2 metres (6 feet 8 inches) or so, and encompasses both some of the rarest as well as some of the commonest of our plant species. A member of another sedge genus is the SAW SEDGE (*Cladium mariscus*). This forms very pure and dense communities at Wicken Fen where it has been regularly cut for thatching and kindling for several hundred years, but is now very much reduced in its distribution elsewhere owing to the drainage and reclamation of land for farming.

Its leaves, which reach up to 3 metres (10 feet) in length, are strap-shaped and evergreen and armed with saw-teeth all around the edge. When they drop off the plant, after a life of two or three years, they are slow to rot, and form a layer with the appearance of 'a thick elastic mattress'. If this community is cut more often than once in four years, the plants are enfeebled and PURPLE MOOR-GRASS (*Molinia caerulea*) enters the scene to share dominance with the sedge; annual cutting rules out *Cladium* completely. Especially after a cutting several damp woodland species may be found including the creeping MARSH PENNYWORT (*Hydrocotyle vulgaris*) with its little round leaves, and the white-

Marsh marigold (Caltham palustris)*: though they can be found growing in any moist place, the finest marsh marigolds are always found actually growing in water. Because of this, the plant is called 'drunkards' in the west of England. The shiny golden cup of the flower is formed not of petals, which are absent, but of five, much-enlarged sepals.*

flowered MILK PARSLEY (*Peucedanum palustre*) known locally as the carrot by fenmen, and the food plant for the swallow-tail butterfly larvae. Where *Molinia* alone is dominant, further flowers can be found including meadowsweet, the tall and striking MEADOW-RUE (*Thalictrum flavum*) with its yellow flowers made bright by their twenty-four spreading stamens, DEVIL'S BIT (*Succisa pratensis*) and the smaller VALERIAN (*Valeriana dioica*).

Other fen flowers include MARSH MARIGOLD (*Caltha palustris*), also known as 'may blobs' (the tight flower buds of which are sometimes pickled and used as a substitute for capers), MARSH CINQUEFOIL (*Potentilla palustris*) which has big purplish flowers, MARSH HELLEBORINE (*Epipactis palustris*) with its beautiful reddish-veined flowers, and the PURPLE and YELLOW LOOSESTRIFES (*Lythrum salicaria* and *Lysimachia vulgaris*). The MARSH PEA (*Lathyrus palustris*) is especially typical of fens. It twines amongst the sedges and grasses with the tendrils on the ends of its leaves, and bears beautiful pale purplish-blue flowers in July. One MARSH ORCHID (*Orchis strictifolia*) is widespread in fenlands, with its flesh-coloured blooms, whilst another, the FEN

Bogland Flowers (from left to right): Bog asphodel (Narthecium ossifragum)*; Butterwort* (Pinguicula vulgaris)*; Bladderwort* (Utricularia)*; Sundew* (Drosera rotundifolia)*; Cotton-grass* (Eriophorum).

ORCHID (*Liparis loeselii*), once abundant, is now very rare, probably from the combination of fen drainage and, strangely enough, the absence of man's activities within them. Variation in species within fen communities is in fact as much to do with our management of them as with local conditions; the richest fen flora arises in the most disturbed areas, and cutting, burning, grazing and peat-digging all contribute to the overall pattern of fen vegetation.

Bogs have a vegetation typically composed of various species of BOG-MOSS (*Sphagnum*), ranging from the hummock-forming to the pool-dwelling, often forming a close carpet but including some associated characteristic plants, with members of the heath family (Ericaceae) on the drier parts. *Sphagnum* is a curious plant. Its pale, clear-green or greenish-yellow branches are covered in little leaves, each of which has a great number of water-storing cells with pores, giving the plant an absorbancy power ten to twenty times greater than cotton-wool.

Though the number of species of flowering plants is small, bogs still manage to contain some of the most fascinating members of our flora. Take, for instance, the BUTTERWORT (*Pinguicula vulgaris*). This strange and

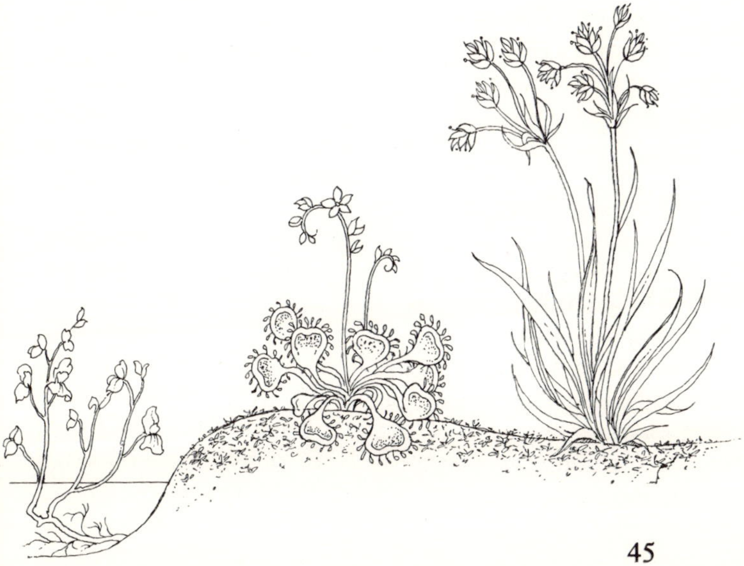

beautiful plant has a few light-yellow fleshy leaves arranged on the ground like the arms of a star-fish, which catch small insects with their sticky upper surfaces and digest them there and then. From the centre of the leaf rosette grow purplish, slender stems, each bearing a single purple voilet-shaped flower.

The BLADDERWORTS (*Utricularia* spp.) have similarly beautiful flowers, this time yellow and double-lipped, borne on the end of a slender stalk which pushes the flowers clear of the water surface whilst the rest of the plant floats underwater in bog pools. The thread-like leaves bear tiny pouches or 'bladders' armed with sensitive hairs which, when triggered, act to engulf small swimming creatures – so fast that the action cannot be followed by the eye or cine camera – and the victims are then digested by the plant.

Another bog plant with an appetite for insects (a handy source of food in nutrient-poor bogs) is the curious SUNDEW (*Drosera rotundifolia*), often to be found growing in tussocks of bog-moss. It has a neat ring of reddish, spoon-shaped leaves with long glandular hairs studded with glistening drops of sticky secretion which attract small insects. An insect alighting on one of these leaves will soon find itself caught in a deadly trap of hairs; digestive juices are then poured over it, enabling the plant to absorb its meaty food.

Rather more common-place members of the bog community are the COTTON-GRASSES (*Eriophorum* spp.), so-called because of their striking white woolly fruiting-heads, for the sepals and petals of each tiny cotton-grass flower together form long white strands which stick out beneath the tiny seed-like fruits. The plants are thus well-adapted for wind dispersal, and in order to facilitate this their dancing heads are held up high above the *Sphagnum* carpet on long stalks to shake in the bog-land breezes. Here too we may find a member of the lily family, the beautiful BOG ASPHODEL (*Narthecium ossifragum*), its star-like, orange-yellow flowers vivid against the greens of the bog-moss. The petals do not fall off but bend upwards to surround the fruit, and the plant may cover quite large patches as it has an underground creeping stem.

7: Flowers of ponds and streams

PONDS AND STREAMS HARBOUR SOME OF OUR FINEST WILD flowers; picture the reflective peace of a calm stretch of water bearing white and yellow water-lilies, flanked by the upstanding yellow flowers of the wild iris amongst the green swords of its leaves, with here and there, perhaps, the pale mauve spikes of the purple loosestrife. Such plants, living as they do submerged or partly submerged in water, are called *aquatics*, and though this term covers a variety of life-forms from minute algae, through mosses and ferns, the flowering plants comprise the bulk of species. We know that life began in water, and that during the course of evolution, plants became adapted for life on land, and so we must consider all our aquatic flowering plants as secondarily adapted for their present mode of living; they have, if you like, gone back to the water.

Here, of course, the problem of drought is over-ruled, and, since water is of a similar density to plant tissue, supportive tissues can largely be done away with, but a different set of problems must be faced. Submerged plants, at least, are dependent upon what is dissolved in the water in the way of oxygen and nutrients. Turbulent streams will obviously contain plenty of oxygen, but it is a fact that this essential gas diffuses more than a thousand times more slowly in water than in air, and roots may find themselves operating under severely limited and limiting conditions. In an effort to counteract this, many aquatics have cavities within their tissues where oxygen released in photosynthesis can be stored, and a network of them provides a system of easy-access routes, as in the PONDWEEDS (*Potamogeton* spp.). Turbulence too in the watery habitat has resulted in the adoption by aquatic plants of a number of characteristic forms aimed at reducing the chances of damage to these typically soft-stemmed plants. Nutrient availability is of course of prime importance in governing what vegetational cover can be supported where, for example, a swift mountain stream running over insoluble rocks will have little chance of accumulating silt of any quality and its flora will be correspondingly poor, whereas a rich aquatic vegetation will be found along silted-up sluggish rivers and ponds. At the other extreme, a pond heavily used by cattle or a stream densely overhung and choked with dead leaves will harbour little vegetation.

A rich aquatic vegetation will, in the course of time, be invaded by marsh and fen species as the silt level rises above the water surface, and a pond exhibits zonation of the species characteristic in this aquatic succession from water to land vegetation. In the centre, where the water is relatively deep, are found the sub-

merged forms of flowering water plants, forms which show the most profound adaptation to this mode of life. Typical of this region are the CANADIAN PONDWEED (*Elodea canadensis*), the WATER MILFOILS (*Myriophyllum*) certain WATER CROWFOOTS, such as *Ranunculus fluitans* and certain pondwoods. The typical shape of these plants can be seen by looking down from the parapet of a bridge. Long, thin and thread-like leaves are borne on flexible stems [the leaves of *Potamogeton compressus* are 10–20 centimetres (4–8 inches) long by only 2–4 millimetres (about $\frac{1}{8}$ inch) wide!] allowing the plant to stream and swirl with the currents. This brings the majority of leaf cells into the closest possible proximity with the water and its dissolved gases and minerals. Many species, like the water crowfoots, have leaves split into thread-like branches ('dissected') as well.

Most submerged plants are rooted on the floor of the pond, but a rather exceptional plant, the HORNWORT (*Ceratophyllum*), has no roots and will float freely just beneath the water surface. It is also peculiar in that,

Pond flowers (from left to right): crowfoot (Ranunculus fluitans)*; water milfoil* (Myriophyllum)*; duckweed* (Lemna)*; the large free floaters – water soldier* (Stratiotes), *frog-bit* (Hydrocharis)*; the water-lilies, white and yellow; half-hearted submergents – arrowhead* (Sagittaria), *and water plantain* (Alisma plantago-aquatica)*; the reedswamp members –* (Typha, Phragmites, Schoenoplectus)*; bogbean* (Menyanthes trifoliata).

unlike most aquatics which share with land plants flowers adapted for development in the air, flowers are borne in this submerged position. Ripe anthers are detached from the male flowers and surface by means of a little float and the released pollen then sinks to hopefully find contact with the protruding stigmas of the female flowers. The depth at which submerged plants can live is limited by light penetration; silty, shady ponds often have a limit of 1 metre (3 feet 4 inches) or so, whilst other, clearer, more open bodies of water will contain species up to 10 metres (34 feet) down.

In shallower reaches, the half-hearted submergents can be found like the ARROWHEAD (*Sagittaria sagittifolia*), named after its three-pointed upper leaves displayed to the air, whilst the lower leaves stream like ribbons in the water beneath, and the WATER PLANTAIN (*Alisma plantago-aquatica*) with its half-metre (1 foot 8 inches) high pretty, finely branched cluster of star-like flowers.

More characteristic of shallower and, most particularly, of sheltered regions, are the floating aquatics. Here the leaf form is usually oval or near-circular, a form thought to minimize wave and wind damage, and it is beautifully exemplified by the WHITE and YELLOW WATER-LILIES (*Nymphaea alba* and *Nuphar lutea*). Few plants can be more beautiful than the *Nymphaea*, the largest of all our wild flowers, with its great white blooms. *Nuphar* too, with its flagon-shaped seed-vessel and spirit-scented nectar (which accounts for its nickname, 'brandy-bottle') enhances many a still pond with its yellow blooms. Another yellow-flowered aquatic with very similar round, floating leaves but not at all related to the water-lilies is the FRINGED WATER-LILY (*Nymphoides peltatum*), which is locally abundant in the Fens. In the Amazon this adaptive form is carried to extremes; the GIANT WATER-LILY (*Victoria Amazonica*) can have leaves up to 1.2–2 metres (4 feet–6 feet 6 inches) in diameter, and is buoyant enough to support a young child!

In this country, other common 'floaters' are the pondweeds *Potamogeton natans*, and *P. gramineus* (which also has submerged leaves). The above species are all rooted, with long trailing stalks so that a rise in water level will not result in 'drowning'. Free-floating species are common too, these being especially abundant on nutrient-rich small ponds and sluggish rivers. DUCKWEED (*Lemna*) is a classic here. Though tiny, this is in fact a flowering plant, consisting of a minute floating, circular plate with a single root descending a millimetre ($\frac{1}{25}$ inch) or so into the water (*L. minor*), or several such rootlets (*L. polyrhiza*), or even none as in the case of *Wolffia arrhiza*, the world's smallest flowering plant, groups of which appear like rafts of green pinheads. Larger free-floaters include the FROG-BIT (*Hydrocharis*), and the WATER SOLDIER (*Stratiotes*), the latter having saw-edged leaves so that it looks like a floating pineapple-top. Winter finds this plant sunk and buried in mud under a heavy incrustation of lime, rising again in the spring when new leaves are produced and the plant once more becomes buoyant.

Since fully aquatic plants must expend a lot of energy

Flowering rush (Butomus umbellatus)*: a beautiful member of the reedswamp zone, the flowering rush has narrow, sword-like leaves as long as its stem. It bears a cluster of pink flowers on the end of a tall unbranched stalk in late summer.*

in bearing and maintaining their flowers above the water, vegetative budding is important in maintaining the species, indeed a rapid spread can result this way. Some plants, like the SLENDER SPIKE-RUSH (*Eleocharis acicularis*), will only flower if the plant is left high and dry by a drop in water level. Where flowers are formed, they are often reduced in complexity; the male flower of *Wolffia* has just one stamen!

The margins of slow-moving rivers or ponds harbour the third zone of aquatic forms, the reedswamp community, standing upright with their roots, stem bases and young shoots in the water. With their typically long flexible stems, they can sway with the wave action, and often provide sheltered pockets of still water for some of the 'floaters' mentioned above. Commonly occurring species include the 'bulrush' or REEDMACE (*Typha*), the common reed, which is in fact a grass, and the LAKE REED (*Schoenoplectus lacustris*). Often these three occur together, zoned in the above order as approached from the shore. Interestingly, the lake reed, in common with another reedswamp plant, the BUR-REED (*Sparganium ramosum*) which has fruits of prickly green burs like little chestnuts, can extend its habitat away from the margins by adopting a submerged form with elongated, flexible strap-like leaves, a form which, however, never flowers. An especially beautiful flower, lending the reedswamp zone a shimmering beauty, is the BOGBEAN (*Menyanthes trifoliata*). A member of the gentian family, it has flower-spikes of foamy white and pale pink, and was much sought-after in northern parts as a herbal remedy for feverish colds. HAIRY WILLOW-HERB, or 'codlins and cream' (*Epilobium hirsutum*), is another attractive flower which commonly forms colonies around streams and ponds. Its leaves are high in vitamin C (140 milligrams in 100 grams of vegetation 0.14% by weight), and its roots were said to contain the power to tame any wild animal! Then, too, there are the pink buds and sky-blue flowers of the WATER FORGET-ME-NOT (*Myosotis palustris*) which grows in profusion along river banks. Another favourite is the SWEET FLAG (*Acorus calamus*), all parts of which have a fragrant sweet scent and its use in early days for covering floors

Water forget-me-not (Myosotis scorpiodes)*: this pretty plant is sometimes called 'scorpion grass' because its flowers are borne in long clusters resembling scorpions' tails because they curl over at the tips, thus protecting the tiny, youngest buds. The forget-me-not leaf is shaped like a mouse's ear, and it is to this distinction that the plant owes its generic name of Myosotis.*

was extensive. It produces a thick, blunt spike of brownish-gold flowers, and tends to be rather overshadowed by surrounding reeds and bulrushes, as are the many other semi-aquatic flowering plants to be found here.

8: Flowers of heathland and grassland

THE SANDY HEATHS AND COMMONS OF LOWLAND ENGLAND and the upland moors which form such typical landscapes in more northerly parts, both bear a similar flora, a flora rather poor in number of species as compared to the bright multitudes found on chalk downland.

Heath and moor plants often dominate large areas of their habitats, as with HEATHER (*Calluna*) and on slightly damper upland soils, the purple moor-grass and the MAT-GRASS (*Nardus stricta*). On badly drained parts 'cotton-grass moor' can develop, types of which grade into boglands. Wild *Calluna*, a small shrub favouring acid soils, has pale purple flowers renowned for their appearance at the end of the summer holidays and also for their dark honey. Its evergreen leaves are adapted to its nutrient-poor home since annual shedding would mean a loss of valuable minerals, while its roots are habitually infested by, and exist in association with, a microscopic fungus (*Phoma*) capable of fixing nitrogen from the atmosphere. Much heather moorland below the tree-limit is in fact capable of supporting oak woodland, but is maintained as such by 'cultivation' in the interests of grouse-rearing, areas being burnt on a rotary system to provide short young shoots for feeding and taller, older clumps for nesting. Heather, with its underground rhizomes, can stand this treatment; most other plants cannot.

Commonly accompanying *Calluna*, especially on drier ground, is the purple BELL-HEATHER (*Erica cinerea*) with its more reddish-coloured flowers and bristle-like leaves, whilst on wetter parts, the CROSS-

LEAVED HEATH (*Erica tetralix*) can be found. The latter bears waxy pink flowers, inside which live tiny insects called thrips, which cause pollination. BILBERRY (*Vaccinium myrtillus*), too, may accompany the heather.

Much moorland is used for grazing, and this the heather cannot stand; intense grazing of heather moor leads to its development into a rather poor-quality grassland in which the SHEEP'S FESCUE (*Festuca oviva*) and species of BENT (*Agrostis*) dominate. Such grasslands frequently harbour relic species of former woodlands like the bluebell, wood anemone and wood-sorrel, hidden beneath the shade of bracken fronds.

Upland heaths, then, originated from, and are maintained by, clearing and grazing. Lowland heaths are periodically burnt over by dry-summer fires and are rabbit-grazed in addition, whilst coastal heaths are

Heather (Calluna vulgaris)*: hundreds of different forms of this plant have been produced by man. They include forms with double flowers, which range from white to rich purple; some varieties can reach a metre (40 inches) in height. (Wild Heather rarely exceeds a height of 60 centimetres (24 inches).) 'Heather tea', concocted from the dried flowers, is enjoyed by some folk, and the plant is thought to be of value in the treatment of bladder and kidney troubles.*

Broom (Sarothamnus scoparius)*: this is a widespread heath plant, though it dislikes chalky soil. It flowers in June and July; the seed-pods are black. Broom tops are used as a diuretic.*

subject to strong winds and salt spray and this prevents the growth of tall woody plants here. Two plants typical of lowland and coastal heaths are the COMMON GORSE (*Ulex europaeus*) and the BROOM (*Sarothamnus scoparius*), often thought to be one of the most beautiful of British shrubs, with its butterfly-shaped bold yellow flowers and curving pliant pale-green branches, much used in former times for the old-fashioned besoms, or brooms. Both broom and the prickly-spined gorse have much-reduced leaves, their thin green stems taking over instead the process of food manufacture, thus water loss through transpiration is cut down, for roots must battle against their operational inefficiency in these typically oxygen-poor soils and water is hard won. Young gorse shoots are very nutritious, and were used extensively as fodder; even the hard old shoots were used, these being ground up in special mills. April sees the healthy hill-sides covered in sheets of gold when the common gorse comes into flower, and later, on still hot days, their scent, reminiscent of pineapple and almonds, lies heavy on the air. Later still, the air is full of continuous 'popping' noises as the dry pods burst open explosively, scattering the shining brown seeds, each with its little oily orange lump (*caruncle*) which ants love to eat, laboriously dragging back the seeds to their nests in order to do so. In this way dispersal is effected, and accounts for the frequency of gorse bushes along heath footpaths; they are ant highways.

Occupying slightly higher ground on western heaths, is the WESTERN GORSE (*U. gallii*), which forms a low scrub barely a metre (3 feet 4 inches) high in the bent-fescue grasslands, its squat shapes being maintained by wind-burn and sheep-nibbling of the young shoots.

Gorse (Ulex europaeus)*: though gorse flowers are at their best in spring, they may be found in bloom on heaths all the year round – hence the saying 'When gorse is out of flower, kissing is out of season'.*

In unshaded positions in heathland, one of our loveliest and best-known flowers can be found, the HAREBELL (*Campanula rotundifolia*), a plant common on chalk grassland too, as is also the yellow-flowered lady's bedstraw. Here too stonecrop runs riot with its fleshy water-storing leaves – avoided by animals because of their sharp, burning taste – and several species of scentless violet may be found.

Especially characteristic of these dry lowland heaths, though very inconspicuous, are the ephemerals, tiny annuals living for mere weeks in spring-time whilst the sandy soils yet retain some moisture, accompanied by small annual weeds from surrounding arable land: the early HAIR GRASS (*Aira praecox*); a FORGET-ME-NOT (*Myosotis versicolor*) whose flowers change from yellow to blue; the tiny RED STONECROP (*Tillaea muscosa*); Britain's smallest flowering plant the CHAFFWEED (*Centunculus minimus*), and many, many more.

Chalk grassland, as typified by Box Hill, the South Downs and Salisbury Plain, forms a 'very English' part of our countryside with its springy turf dotted with bright flowers. Traditionally used as sheep pasture, the plant shoots may be further reduced by rabbit-grazing to rise little more than 2 centimetres ($\frac{3}{4}$ inch) from the dry soil, and here characteristic plant associations form. Typically dominant is the sheep's fescue; other common grass species include bent, and in less-grazed parts, the silvery-spiked *KOELERIA GRACILIS*, UPRIGHT BROME (*Zerna erecta*), and the yellowish HEATH FALSE-BROME (*Brachypodium pinnatum*), the latter spreading widely in some areas due to its unpalatability to rabbits. Grasses, generally, are able to withstand grazing as their stems and buds lie close to the ground, and are thus not usually damaged; the process in fact ensures constant leaf re-growth. BIRDS' FOOT-TREFOIL (*Lotus corniculatus*), or 'bacon-and-eggs' because of its red-streaked yellow flowers, a widespread plant of open habitats, flourishes and spreads in heavily grazed turf too, since it produces slender creeping, and occasionally rooting, stems. This plant also has a long tap-root capable of drawing water from a considerable depth, a definite advantage in these dry-surface soils. More con-

Common rock-rose (Helianthemum chamaecistus)*: this flower is visited by insects not for its nectar, but for its pollen, which is taken as food. The vitamin content of hive-bee honey in fact derives from the pollen grains which it contains.*

fined to chalk grassland is another member of the pea family, the locally abundant HORSESHOE VETCH (*Hippocrepis comosa*) bearing golden flowers, later to be replaced by curved pods splitting into horseshoe segments. COMMON ROCK-ROSE (*Helianthemum chamaecistus*) is familiar here, favouring trampled areas especially, and a plant often associated with it is the pretty, trailing WILD THYME (*Thymus* spp.), 'the close-bit thyme that smells like dawn in paradise'. Other common associates include SALAD BURNET (*Poterium sanguisorba*) with its feathery leaves smelling of cucumbers and roots up to a metre (3 feet 4 inches) long, STEMLESS THISTLE (*Cirsium acaule*) with its single dark-crimson flower-head, several members of the carrot family, most notably BURNET SAXIFRAGE (*Pimpinella saxifraga*) and many weeds common on arable land.

Man orchid (Aceras anthropophorum)*: the green sepals of this orchid flower form a hood over the yellowish-petal lip. Orchids have a fascinating device for the safe carriage of their pollen from one flower to another. Instead of powder, their pollen is stuck together in a pair of club-shaped balls which have short stalks with sticky bases. (An orchid, in effect, 'puts all its eggs in one basket'.) A visiting bee will leave the flower with a pair of these pollen balls stuck firmly to her head. During the next minute, the stalks of the pollen-balls gradually bend forwards. In this position, the pollen-balls are 'all lined-up' for perfect contact with the stigma lobes of the next orchid flower that the bee visits.*

Several orchids occur here too including, though rather uncommonly, one with a drooping flower lip in the form of a human figure!

More typical of northern limestone grassland are LADY'S MANTLE (*Alchemilla*), BLUE MOOR-GRASS (*Sesleria caerulea*), an indicator of alkaline soil, and the beautiful MOUNTAIN PANSY (*Viola lutea*) whose flowers vary from yellow to purple. The sharp cliffs and crags ('scars') of these northern grasslands are a haven for some rare flowers; several, like the HOARY ROCK-ROSE (*Helianthemum canum*), are relics from the naturally treeless vegetation of ten to fifteen thousand years ago, escaping here the relentless advance of the forest, with its 'adapt or perish' implications.

9: The history of our vegetation

IT WAS NOTED IN THE INTRODUCTION THAT THE FLOWERING plants did not exist at the time when the great coal deposits were being formed (the so-called *Carboniferous* period). Although plants began to live on land around 400 million years ago, the angiosperms first appeared a mere 150 million years ago; they are essentially a young group, the evolution and spread of which must have been very rapid to attain their present diverse and wide-ranging state. Though there are many ingenious mechanisms associated with plant dispersal, mechanisms which may be seen to be very efficient over relatively short distances, it seems unlikely that these can explain the basic similarities which exist between the components of many floras from different parts of the world. These areas are sometimes separated by vast oceans, and though former land-bridges undoubtedly played a rôle in the rapid distribution of species, the theory of *continental drift* seems much more helpful here. It seems that the continents were closer together in the past, thus allowing the spread of evolving species. Britain occupied a position much further south than at present, and was supporting a tropical swamp-forest with a rich flora approximately fifty million years ago. As the continents gradually drifted to their present position, our flora changed to a north temperate one.

Twenty thousand years ago, the Earth became very cold. This heralded the era known as the 'Ice Age'. (Though, perhaps more properly, it should be the 'Ice Ages', for there were actually three, with warmer periods in between.) Huge sheets of ice covered Northern

Europe, including Britain except for an area south of a line from the Wash to the Bristol Channel. Britain was at this time still joined to the continent by a land 'bridge' which now lies submerged beneath the North Sea. (Into this bridge, flowing northwards, ran the great river Rhine, with the Thames as one of its tributaries.) Around 15000 BC, the last of these ice sheets retreated, leaving a raw, fertile soil into which species from the ice margins further south were able to migrate.

In this 'Late Glacial' period, a rich, but open, treeless flora developed in Britain. (A vegetation of this sort is sometimes called '*park tundra*'.) The chief plants were arctic-alpines like the mountain avens, a species which became particularly widespread, and DWARF BIRCH (*Betula nana*), amongst clumps of sedges and grasses and scattered carpets of mosses and lichen. The disappearing ice left a very waterlogged land in its wake. Lakes and marshes lay everywhere in the lower parts, and flowers characteristic of these habitats, like the pondweeds and the fragrant meadowsweet, flourished. So, too, on the drier parts, did many of our present-day weeds like the DOCKS (*Rumex* sp.) and GOOSEFOOTS (*Chenopodium* sp.), as well as a variety of species of chalk and limestone grasslands, a number of which are now much rarer. This glacier-scoured terrain, then, supported a rich plant cover, and this in turn provided food for populations of large grazing wild animals, like the elk, which was hunted by early man in Britain.

It is from this Late Glacial period that the vegetational cover of Britain really dates. Since a number of plants did not succeed in returning to this country during the period following the Ice Age and before we became cut off from the continent, our flora does not contain as many species as some other countries. Tropical areas, for instance, were unaffected by the Ice Age, and have ancient floras which can be measured in millions rather than thousands of years. These areas harbour a truly bewildering variety of plants.

The history of how our flora has developed in the fascinating period since the Ice Age, has been gathered with the aid of two important processes; these are *pollen analysis* and *carbon dating*. Peat in bogs and lakes is formed of layer upon layer of plant remains, the oldest deposits lying at the lowest levels, and the youngest at the top. A long bore taken through an ancient accumulation of peat will contain samples of these successive layers and caught up within each, in an often amazingly well-preserved state, is an assortment of pollen grains.

Pollen grains are able to resist decomposition because they have very tough coats; furthermore, different plants have differently shaped grains bearing characteristic sculpturings which makes them quite easy to recognize under the microscope, after they have been treated in special ways. The pollen is especially representative of wind-pollinated species like the grasses and trees, if there are any, growing in the countryside surrounding the peat bog. Tree pollen is wonderfully adapted for wind dispersal, being minute and produced in profusion, well above ground level; the relative proportions of the particular species in a sample give a good indication of the composition of the forests in the vicinity. Overall comparisons of samples can show how the vegetational cover changes in time.

The dating of peat samples from which the pollen is analysed, can now be determined very accurately by using the *radiocarbon* or 'C14' method. A proportion of the carbon in the atmosphere consists of radioactive carbon 14, and living organisms incorporate this proportion into their tissues during life. With the passage of time, the radioactive material gradually decays at a fixed rate, and so, by determining the proportion of radiocarbon still residing, the age of the organic remains can be measured. Evidence from other sources is fitted into the picture too, archaeological finds being especially informative concerning the period when man first began to make his mark upon the land, and in so doing left behind fragmented records of his all-but-forgotten life.

The Late Glacial period was a very cold one. Average summer temperatures were 4–5 degrees C (8–9 degrees F) lower than at present, but conditions gradually began to improve, and, as they did so, trees crept into the open terrain. Thin forests of HAIRY BIRCH (*Betula pubescens*), JUNIPER (*Juniperus communis*) and ASPEN

(*Populus tremula*) developed first, and WILLOWS (*Salix* sp.) grew around the lakes. Around 6000 BC, the climate had become definitely warm and dry, and forests of PINE (*Pinus sylvestris*) dominated the scene, along with abundant hazel. Pollen samples from around this period show a distinct drop in the proportion of herbaceous pollen; the herb community had been ousted from its previous dominant position by the advancing forests.

These forests were inhabited by man at a stage of culture known as *Mesolithic*. He had flint tools and lived by hunting and collecting wild plants (the so-called 'hunter-gatherers') and as such made little impression on his environment. His numbers were low since a large area of natural vegetation was needed to support just one person. (Modern hunter-gatherers, the South American native Indians, need a hunting area of over 26 square kilometres (10 square miles) for one human.)

Meanwhile, Britain was sinking. By the end of this warm, dry spell, the eastern parts were 30 metres (100 feet) lower than at the beginning, and the land bridge had been engulfed by the sea. This happened in 5500 BC, a memorable point in time, for a *native* species is defined as one for which there is evidence for its occurrence before this time. Species arriving after 5500 BC are assumed to have resulted from the purposeful or accidental activities of man.

Possibly arising out of its new status as an island, Britain now experienced a dramatic turn-about in its weather; it became very much wetter. The pine forests fell back, and broad-leaved deciduous trees, more suited to this climate, came into the picture in force. Chief amongst these was the OAK (*Quercus*) which formed great forests throughout most parts of the land.

It was as this mild, wet, 'Atlantic' period was drawing to its close, around 3000 BC, that certain farmer immigrants infiltrated amongst Mesolithic man. They heralded the start of the 'Neolithic revolution' with their practices of organized food production, which were to have such a profound effect on the vegetation of Britain.

10: Man and nature

FLINT TOOLS HAD BEEN USED BY MESOLITHIC MAN, BUT the Neolithic farmers developed a characteristic technique of fine polishing of their implements. With their 'new, improved' stone axes, these people were able to tackle the task of forest clearance to practise their 'shifting cultivation'. A cleared plot was burnt and cultivated by sowing the seeds of many varieties of fruitful plants ('crops'), until the soil was exhausted, whereupon it was abandoned, and gradually became closed in again with regenerated forest. One such axe-head recently felled over a hundred trees, though it had not been sharpened for 4,000 years! Three men were able to clear, in this manner, 500 square metres (600 square yards) of SILVER BIRCH (*Betula pendula*) forest in four hours. Pollen samples covering this period show a great increase in birch pollen after the burning stage, for birch seeds need shade-free conditions in order to germinate. At the same time, pollen of herbaceous plants, mixed with that of pasture grasses and cereals, jumps to the fore. Amongst a number of weeds, RIBWORT (*Plantago lanceolata*) was especially abundant.

Though the significance of this was slight to begin with, the scale upon which it was practised was ever-increasing, so that eventually the non-tree pollen rivalled that of the forest trees; now the people could keep pace with their population increase by clearing more land to plant more seed and graze more animals. The demand for axe-heads for forest clearance became so great that Neolithic man had to set to with red deer antlers and the shoulder-blades of cattle to mine flint from underground seams!

Following a drier interlude, in which BEECH (*Fagus sylvatica* ('beech' is derived from an Anglo-Saxon word 'boece' and old Teutonic 'boka', indicating its edible fruit)) became increasingly abundant in Britain, and Bronze Age man was widespread, the weather turned wet and much cooler, a climate which has persisted up to the present day. Fed by this high rainfall, two characteristic types of bog formed. Where the level of fen peat, accumulating round the edges of the numerous small lowland lakes, rose above and away from the effect of its alkaline ground water, it developed into *raised bog* (because it appears like a slight lump upon the surrounding fen), bearing the acid vegetation, bog-moss especially, so typical of bogs. In England many of these raised bogs have since been drained and the peat burned for fuel (Mediaeval peat cutting in the Norfolk Broads removed about 280,000 cubic metres (900,000,000 cubic feet) of peat!), but they can still be seen in Ireland. Bogs with a similar kind of peat, and an assortment of plants, came to form in the wettest regions of all; the west coasts of Scotland and Ireland, and on badly drained uplands. They covered the ground in a continuous thick layer (*blanket bogs*).

In the latter part of the Iron Age, not surprisingly, man chose the drier parts for his settlements. Like Neolithic man, he favoured the chalk uplands of the south-east, and built forts and camps on other hill-top sites. But the extent of his influence was further reaching. Wielding iron axes, he was able to lay waste vast tracts of the wild oak forest where lynx, bear and wolf still roamed, and which still covered much of the fertile, but heavy-soiled, British lowlands. In so doing, he further extended the areas of cultivation which had been started on the drier, lighter soils in the south-east and on the chalk uplands.

Anglo-Saxon Britain saw a countryside of scattered forests, heaths, grassy pastures and arable fields. These fields were not enclosed, being community-owned and cultivated in common by the villagers; hedgerows were not in evidence yet. Domestic animals were frequently grazed in the forests where the villagers also collected their firewood. Though pigs may have helped to regenerate oak and beech forests by trampling and nose-churning of acorns and beechnuts into the soil, and by reducing the numbers of small nut- and seedling-eaters, like voles, the overall effect of this grazing was the destruction of large areas of forest; the many wild flower species of the forest field layer were replaced by grasses. Not only was the soil compacted, and young seedling trees destroyed by the stamping feet, but the young buds and shoots of older trees were removed up to the reach of the tallest browsers. Heavily grazed areas gradually became poor grassland as the old trees died and were not replaced, whilst less-intensely grazed parts developed into rough heaths composed of plants which could protect themselves from attack by their thorny nature, like gorse and SLOE, or BLACKTHORN (*Prunus spinosa*). Such areas were named, rather aptly, forest 'wastes'.

In 1066, the Normans arrived. They created the 'Royal Forests', within which cultivation was prohibited, primarily to conserve deer, but also the trees. By the twelfth century, these forests covered almost a third of the country! Later the laws were relaxed when it was realized that an ever-increasing population demanded more space for food production and cutting-back of the Royal Forests began. From this period onwards throughout the Middle Ages, land for agriculture and pasture was increasingly brought into existence at the expense of the natural forest communities.

The Mediaeval period, in particular, saw a great increase in pasture land, for wool became a chief source of industry; in the fourteenth century, there were four sheep per head of the human population! Corn, too, proved profitable and was grown on a large scale. However, wood still remained a vital part of the economy. Timber was needed for building houses, bridges, mills and construction generally; brushwood became extensively used for fencing in crops, to protect them from grazing animals. Large quantities were needed for domestic fuel in addition to iron-smelting, and the glass, gunpowder and brick manufacturing industries. Oak bark, too, was valuable as it was widely used for tanning hides. Overall this led to a great depletion in our natural timber reserves. 'Disafforestation' was such that by the seventeenth century, the Royal Navy was forced to import wood. Perhaps this is hardly surprising for in order to build just one of their great ships, more than two thousand big oak trees were needed!

It was soon realized that timber was a valuable, not easily renewable resource, and widespread protective practices were called into operation. *Coppicing* (from the French word 'couper', to cut involved cutting a plot of woodland, and enclosing it for about seven years to recover undisturbed. Sometimes widely spaced '*standards*', usually oaks, were left to grow wide girths [one oak in Dorset was measured at 20.4 metres (67 feet) round!] and big spreading side branches yielding curving timber valuable in shipbuilding. 'Coppice-with-standards' is still the most common type of woodland to be found in the Midlands and southern regions, with its renowned springtime field layer of wild flowers. Today coppicing is generally unprofitable and copses are maintained more for pheasants and foxes; coal and later oil replaced wood as a fuel, imported bark and extracts were used instead of oak bark for tanning, whilst the old brushwood fences (which needed frequent replacement) were superseded by living hedges of HAWTHORN (*Crataegus monogyna*), banks or stone walls, which were to become such a typical part of the British countryside.

Apart from coppicing, it became fashionable in the twelfth century for land-owners to plant trees, and prizes were even offered to those who could plant the largest numbers in a specified time! Planting was carried on into the nineteenth century, but as the trees were privately owned and offered such a slow return on capital outlay, it never met the demands of the country. Consequently, in 1919, the Forestry Commission was set up to provide wood for the nation. They replanted cleared ground as well as existing woodlands, using mostly exotic conifers because of their faster growth rate and the big demand for 'softwood' products, particularly paper. The Forestry Commission is currently planting at a rate of about 24,000 hectares

(60,000 acres) a year, in the proportion of one area of hardwoods to ten of conifers. Privately owned woods cover around a twentieth of the land surface and though, again, many contain exotic specimens, the thoughtful care with which they are planted and maintained can provide a beautiful and varied landscape.

From the seventeenth century onwards, the proportion of cleared land left to pasture increased, especially during the nineteenth century when agriculture was abandoned in favour of the industrial towns. By the start of the Second World War, grass fields covered over half the land. Later the balance shifted in response to the great demand for home-produced food, as the land was put to use in earnest, thus marking another notable period in which our native flora suffered a set-back.

The last couple of centuries have also seen the uprising of a widespread consciousness of the value and appeal of the countryside in aesthetic terms. In 1895, the National Trust for Places of Historic Interest and Natural Beauty was established, and many nature reserves were to follow. This has happened not a moment too soon, for with the great expansion of industries and suburban housing [at the moment, almost 40,000 hectares (100,000 acres) per year are lost to development of all kinds], allied with the widespread adoption of modern intensive farming techniques, including drainage, ditch and hedge removal, and use of herbicides, the remaining natural habitats are threatened on all sides; the overall trend, started in Neolithic times, towards a reduction in the area covered by natural vegetation as a result of man's activities, continues. But it is also true to say that man has reduced the 'monotony' of it too, for he has created many, beautifully varied 'artificial' habitats, like the coppiced oak woods, the sheep-grazed downland, the hedgerow and reservoirs, which have resulted in the richly diverse flora of today. Man is clearly a potent ecological force, but his powers of reason and research must not be forgotten either. In this lies hope for the achievement of dynamic harmony between man and nature.

11: Our changing flora

APART FROM THE BIRD'S-EYE-VIEW CHANGES IN THE main plant communities over the years, changes on a more specific level have occurred both within plant communities and in their distribution. Man has introduced exotic species, and controlled existing ones to his own needs, in the interests of food crops, game and forestry. With the population density and mobility of the British today, our flora cannot remain constant; there are losses and gains. The gains, however, are greater than the losses; since the sixteenth century, when the first botanical records began, only twenty native species have been lost, though many more have recently declined markedly. 'Picking pressure' seems to be responsible for the present scarcity of certain species like the SNOW GENTIAN (*Gentiana nivalis*), the RED ALPINE CATCH-FLY (*Lychnis alpina*), and of showy flowers in bands around towns; habitat destruction by drainage accounts for others, like the FEN ORCHID (*Liparis loeselii*), and by ploughing, as in the case of FIELD SOUTHERNWOOD (*Artemisia campestris*). Yet others show a 'natural' or 'biological' decline, like the CORD-GRASS (*Spartina alterniflora*) which, following its introduction into Southampton Water, failed in competition with its offspring, the hybrid *S.* × *anglica*. Probably the biggest threat to our flora, however, is posed by the changes in arable farming techniques during this century, and in this connection, 'weeds' are very interesting.

Weeds are essentially opportunists, quick to colonize disturbed land with their numerous, often wind-borne seeds, which is why they are so common in gardens, arable land, rubbish-tips and building sites. Many have evolved alongside man from his early days of agriculture, like the FAT HEN (*Chenopodium album*) which, with its grainy seeds, was used as poultry food (the leaves of this plant can be eaten like spinach); others have been introduced, like the annual SPEEDWELL (*Veronica persica*), which came to Britain only about 150 years ago. Some weeds have been introduced, but have still had long associations with man. Such a species is the GOLD OF PLEASURE (*Camelina sativa*), with its oil-yielding seeds, which was a frequent weed amongst flax.

After the last glaciation, there were large areas available for colonization, and probably much of our present weed flora was widespread at this time. As the forests spread in the warmer post-glacial period, the weeds were ousted to marginal places like sand dunes and mountain ledges, and they have only been able to spread back through man's cultivation practices. Today there are around 150 species of plants referred to as 'weeds' by British botanists; this is not really surprising as half of the country is agricultural land. Recent farming practices have, however, reduced their numbers considerably. Liming fields, for instance, has been responsible for a great reduction in the beautiful CORN MARIGOLD (*Chrysanthemum segetum*), once a common sight in the cornfield, with its tall-stalked, brilliant yellow flowers. The little FIELD SPEEDWELL (*Veronica agrestis*) has suffered in this way too. The fragile beauty of the pink-flowered CORN-COCKLE (*Agrostemma githago*) and one of the prettiest British wild flowers, the CORNFLOWER, or 'bluebottle' (*Centaurea cyanus*), have largely disappeared from the scene through seed-cleaning processes (that is, screening out the unwanted seeds). Apart from competing with crop plants, the corn-cockle and another disappearing arable weed, the semi-parasitic FIELD COW-WHEAT (*Melampyrum arvense*), have poisonous seeds which spoilt the flour if they were included in the corn, and so it has been especially necessary to exclude them.

This century has seen the increasing use of herbicides; today almost two-thirds of Britain's arable land is treated annually with these powerful substances. Their uncontrolled use can have catastrophic effects on plant life, for instance, it only takes half a part per million of paraquat to kill most of the aquatic plants in a lake or reservoir. Species which are particularly susceptible to the herbicides commonly applied to cereal crops, include the gay common poppy with its flaming scarlet flowers, the CORN BUTTERCUP (*Ranunculus arvensis*), the trailing purple-flowered COMMON VETCH (*Vicia sativa*) and charlock, or WILD MUSTARD (*Sinapsis arvensis*). Without the appliction of herbicides, a field overgrown with CHARLOCK looks decidedly yellow, almost as if charlock was the cultivated plant! (It is strange to think that a bee, the eye of which covers a different part of the colour spectrum from man's, would see this as a sea of crimson.) Even where herbicides are used, some weed species remain as common as ever, sometimes (because of reduced competition) they are even more common! This can happen with flowers like the little COMMON CHICKWEED (*Stellaria media*), the speedwells, the pretty FIELD FORGET-ME-NOT (*Myosotis arvensis*), BLACK BINDWEED (*Polygonum convolvulus*) and fat hen too. Because weeds are so well adapted to survive long periods of adverse conditions, with lots of hardy, scattered seeds that can stay dormant for years and still remain viable – [the WILD RADISH (*Raphanus raphanistrum*) can produce over two thousand seeds from one plant, and these can still retain the power to germinate after fifty years in the soil!] it seems very unlikely that they will disappear altogether from the countryside, even though their habitats are under pressure. Nature has ensured that they will always be ready to exploit any opportunities that might arise.

With the vast reduction of the native oak forests, and the conifer-biased planting schemes of the Forestry Commission, the deciduous broad-leaf trees of this country have been left sadly impoverished. Consequently their field-layer species, like the anemone, primrose, violet, bluebell, enchanter's nightshade, sanicle and dog's mercury, have all suffered a loss of ground. In the pockets where such woodland remains, however, these flowers are often prolific.

The new, exotic plantations can seem so dense and

dark as to exclude all but the hardiest wild flowers, but time passes, and the trees are thinned, and the bare forest floor rapidly becomes colonized by seedlings. Sometimes foresters use 'wasteland' for their plantations, to the detriment of wetland species like sundews and bog asphodel. The area affected is, however, relatively small, and above 500 metres (1,600 feet) wind exposure prevents afforestation and thus protects these bogs from development.

The hedgerow, though initiated by man, offers one of the richest of habitats. To plants, it is like a continuous ribbon of woodland edge, with the added advantages of extra light and air. A large number of its taller members come from the rose family: hawthorn, sloe, bramble, CRAB APPLE (*Malus sylvestris*) and WILD RASPBERRY (*Rubus idaeus*) are common. Autumn brings a bountiful harvest of wild fruits here, and bird life flourishes accordingly. In summer, the hedgerow is decked with possibly the most beautiful of British wild flowers, the DOG ROSE (*Rosa canina*), with its sweetly scented, pinkish-white flowers, quickly replaced by the familiar scarlet 'hips'. (Superstition has it that the root of this plant cures the bite of a mad dog – hence its name.) Climbers flourish too, like the OLD MAN'S BEARD (*Clematis vitalba*) with its curling leaf stalks becoming hard as wire, and its distinctive 'beard' of fluffy, grey fruits. This plant is also called 'traveller's joy' because it forms natural, shady arbours at the wayside for the hot and weary. Amongst the honeysuckle and the tangled WOODY NIGHTSHADE (*Solanum dulcamara*), the LARGER BINDWEED (*Convolvulus sepium*) supports its elegant pure-white blooms, whilst the TUFTED VETCH (*Vicia cracca*), another climber, adds to the scene with its dangling bluish flower-clusters. Adding lacey ribbons of tiny white flowers to the summer hedgerows are hogweed, or 'cow parsnip', and other relatives from the carrot family. These harbour a rich assortment of insects, many of which are beneficial to man as they pollinate his crops and prey on pests. In spring and early summer, hedgerow banks are studded with the lovely trailing GERMANDER SPEEDWELL (*Veronica chamaedrys*), often called 'bird's-eye' or 'angel's eyes' because

The Hedgerow in Midsummer: including Traveller's joy (Clematis vitalba)*; Dog rose* (Rosa canina)*; Honeysuckle* (Lonicera periclymenum)*; Larger bindweed* (Convolvulus sepium)*; Tufted vetch* (Vicia cracca)*; Woody nightshade* (Solanum dulcamara)*; Cow parsnip* (Heracleum sphondylium) *and relatives; Germander speedwell* (Veronica chamaedrys)*; Greater stitchwort* (Stellana holostea)*; Herb-robert* (Geranium robertianum).

of its bright blue flowers. The dainty pure-white stars to be found here belong to the GREATER STITCHWORT (*Stellaria holostea*). Set on thread-like (hence 'stitch') stalks, the flowers bear ten orange-red stamens, half of which protrude from the flower-tube. Here too, the red-purple flowers and curious beaked fruits of herb-robert can be found. Many other 'marginal', or half-shade, flowers grow in this habitat.

The last twenty years has seen the destruction of over half a million hedgerows, corresponding to about 9,600 kilometres (6,000 miles) each year. At this rate, the last hedge in England will disappear during the first

decades of the next century! Happily the number of rare species threatened with extinction by hedge removal can be counted on one hand. In individual localities, this number can rise to fifty; hopefully these species will be harboured in a neighbouring area.

From the earliest times, the movements of man have affected the flora of Britain. Species like the daisy and GREAT PLANTAIN (*Plantago major*), which can withstand trampling and need freedom from the competition of taller plants, found an extension of their habitat along trackways. Animals driven to distant markets must have conveyed seeds on their bodies, whilst man dug in whips and stakes of willow and poplar, some of which took root, along these drove roads. An interesting weed of waysides and trampled places, which has increased enormously over the years, is the RAYLESS MAYWEED (*Matricaria matricarioides*). Though a relative of the dandelion, this flower lacks the usual special adaptations to wind dispersal typical of its family; it seems that this plant spreads by sticking its seeds to tyres, wheels and boots – a plant hitch-hiker in fact!

Of course road construction has taken its toll of the land; a modern motorway will occupy a swaithe of land around 60 metres (200 feet) wide [about 10 hectares (25 acres) to the mile]. But there is a potential gain here, often overlooked, which may be greater than the loss. Road verges form a continuous open network, allowing the spread of species along them and thus through a country-wide range of soils and differing climatic conditions. In addition, they are little-disturbed (especially along motorways, where pedestrian access is forbidden), and so can provide valuable reservoirs for wild flowers like certain orchids and the 'snake's head' or FRITILLARY (*Fritillaria meleagris* – beautifully named after the flower's dice-box shape and dappled coloration resembling that of a guinea fowl), which are increasingly under pressure from loss of habitat in lowland agricultural and industrial regions. With a total area approaching that of the National Nature Reserves, their value can be seen; verges in fact constitute the richest single habitat for wild plant species in the country. (41% of the total flora of Cambridgeshire has been found along road verges!) Unlike hedges, verges are increasing.

Whereas previously verges were heavily sprayed with chemicals to promote 'tidiness', today herbicide use is carefully retricted, again providing a valuable refuge for certain species. (Perhaps the uniformly depressed verges following on the heels of a universal chemical spraying policy in the U.S.A. served as a lesson in time here.) In addition, botanically important sections of verge are now demarked for special protection. Today motorway verges are planted with trees and shrubs on a scale rivalling that of the eighteenth century, and planners, with restoration rather than decoration at heart, try to create the impression of 'moving unimpeded through the countryside, which comes right up to you on either side, so close that you can almost lean out and touch it'.

Modern transport enables large samples of the world's flora to be brought to the doorstep of Britain, intentionally or otherwise. Some of these *aliens* make themselves at home quite happily, enriching our flora (that is, they become *naturalized*). Mostly, however, they find conditions unsuitable and disappear from the scene again. Waste tips around docks often harbour fascinating, though short-lived, alien introductions.

Whilst man has undoubtedly subjected the natural plant communities to increasing pressure with his increasing population, spread of towns, intense industrial development and pollution, he has at the same time created new opportunities for plants. In this way, he has helped create diversity. Industrial tips, for example, have made conditions where strains of grasses able to withstand high concentrations of toxic metals, such as zinc and copper, have been selected for. Warm water

The Hedgerow in Autumn: similar species to previous page but in fruit: 'beard' of Clematis vitalba*; 'hips' of dog rose; 'haws' of hawthorn etc. plus blackberries on bramble, sloe* (Prunus spinasa), *crab apple* (Malus sylvestris), *seeds of cow parsnip and dock* (Rumex), *beaked fruits of herb-robert, 'clocks' of dandelion etc.*

conditions arising from atomic power stations can create a local diversity of species. Open spaces abound in and around cities through man's activities; these quickly become colonized by hardy species with lots of wind-borne seeds like rosebay willowherb and COLTS-FOOT (*Tussilago farfara*) with its 'parachutes', so light and downy that they are used by goldfinches to line their nests. Old walls and cracks in masonry provide homes for flowers like stonecrop, or 'wall-pepper', and its many relatives, and the WALL PENNYWORT (*Umbilicus rupestris*) with its odd round leaves and drooping greenish-white flowers. In fact, the longest species lists, for a given area of land, are likely to be found actually within an urban area!

Perhaps, after all, change is not such a bad thing. There is an elusive quality to the natural beauty of wild flowers. Perhaps the preceding pages have served to illuminate it momentarily, for, in the words of an old Chinese proverb: *The art of life lies in a constant re-adjustment to our surroundings.*

Index

Honeysuckle
Lonicera periclymenum
This has a strong sweet smell and is found in woodlands and hedgerows from June to September. As its name suggests it is rich in honey and attracts bees, moths and butterflies. It always twines from left to right. It is also known as the Woodbine and is often used as a symbol of love and affection.
Thrift
Armeria maritima
It grows on rocks and cliffs by the sea shore, flowering from April to October. The flowers grow up from tight clumps, like cushions, to six inches in height. It is also known as Sea Pink.
Coltsfoot
Tussilago farfara
It is found mainly on waste ground and in fields where the soil is heavy. It flowers from March to April. The flowers always come first and the leaves afterwards. The Latin name is derived from tussis a cough, because of its reputation as a remedy for lung complaints. The leaves are still used in the manufacture of herbal tobaccos.
Rose-bay Willow-herb
Epilobium angustifolium
Found on banks and in open glades, flowering from July to September, it often grows in a mass up to four feet in height. It is also known as French Willow or Fireweed. Epilobium comes from the Greek, epi upon and lobon pod, because the long narrow seed vessel is crowned by the flower.
Foxglove
Digitalis purpurea
Found in woodland and wasteland, flowering from June to September. it is poisonous, but gives the chemist a drug that is still used in treating heart diseases. The Latin digitalis means finger stall or thimble, from the shape of the flowers; purpurea indicates its colour. It is sometimes known as Finger Flower.
Harebell
Campanula rotundifolia
Found mainly on heaths and chalk downs, it flowers from July to September. Some say that its name should be Ha'erbell as an abbreviation of Heather, where it is often found. It is the Bluebell of Scotland. The Latin name means a little bell.
Sheep's Bit Scabious
Jasione montana
Also called Pin Cushion it is found on heaths, downlands and some meadows, flowering from May to August. It belongs to the campanula family, although at first glance it may seem to be a smaller version of the Field Scabious. The Latin scabies means the itch, as it was thought to cure this complaint.
Wild Strawberry
Fragaria vesca
The tiny fruit is sweet and delicate and found on shady banks and woodland clearings from April to July. Its name comes from the Anglo-Saxon Streowberie, meaning strewn or straying, due to the runners by which it propagates itself. The Latin is fragra fragrant and vesca edible.
Wild Thyme
Thymus serpyllum
It grows on high land usually among close cropped grasses, flowering from June to August. The name comes from the Greek thymus sacrificial smoke, from the practice of burning the plant which gives a sweet smell. The egg of the Large Blue Butterfly is laid singly among its buds.
Marsh Marigold
Caltha palustris
Found by streams, rivers and on marshy ground, it flowers from March to July. The plant is poisonous, but the unopened flower buds are sometimes pickled and used as capers. Known also as King Cup or Marybud and in Scotland as Luckan Gowan, its botanical name is derived from the Greek kalathos a cup and Latin palus a marsh.
Meadow-sweet
Filipendula ulmaria
A fragrant flower found in meadows and beside streams and rivers, on stems two or three feet in height, it flowers from June to September. In Elizabethan times it was strewn over the floors to give a sweet smell. It was said to 'make the heart merry and delight the senses'. Beside the flower you can see the Hoverfly, which is attracted to it by its sweet smell as are many other insects, although it contains no honey. It is known as Queen of the Meadows or Bridewort.
Biting Stonecrop
Sedum acre
It is found on rocks, old walls and other dry sunny places, as its fleshy leaves enable it to store water. It flowers from June to July. Sometimes it is known as Wall-Pepper. The Latin name comes from sedere to sit.
Deadly Nightshade
Atropa belladonna
Found in ruins and stony wastes, it flowers from June to August. The plant has an unpleasant smell and is very poisonous. The berries are especially so and these have often proved fatal. It is also known as Dwale, from the Anglo-Saxon for foolish, and as Devils Cherries. Under the name Atropine or Belladonna it is used in medicine. Elizabethan ladies used the juice in their eyes to dilate the pupils and make them look beautiful — hence the name Belladonna. Atropa is from the Greek Atropos the Fate that cuts the thread of life.
Field Poppy
Papaver rhoeas
Seen in cornfields from June to August on stems one or two feet high, they contain a mildly narcotic juice, used in the middle ages as a remedy for pleurisy and ague. The distinctive seed heads contain many tiny black seeds. The name Pa Pa was given to the poppy by the farmers of Mesopotamia 6,000 years ago.
Sainfoin
Onobrychis viciifolia
Found on chalk hills and downs, it flowers from June to September. It is also grown as a fodder plant, sainfoin being French for wholesome hay. Onobrychis is derived from two Greek words meaning the ass, and to bray.
Viper's Bugloss
Echium vulgare
Growing on chalky ground and sand-dunes up to three feet in height, it flowers from June to September. It is rich in honey and attracts many insects, especially the Spotted Burnet Moth. Its name comes from the Greek bous an ox, and glossa a tongue, from the nature of its leaves which are very rough.
Milk Thistle
Silybum marianum
Found flowering from May to August it grows from three to six feet in height. The intricate white veining on the leaves distinguishes it from all our other thistles. In the past the roots, young leaves and flowerheads were used as food, and grown as a crop.
Toadflax
Linaria vulgaris
Found in hedgerows, fields and waste places, it flowers from July to October. It is also known as Butter and Eggs. The name comes from the Latin linum flax, but it is not a true flax and not in the least useful as such.
Bindweed
Convolvulus arvensis
This climbing creeping plant grows in fields and hedgerows from June to September, twining from right to left. It is loved by bees and long tongued insects. From the Latin convolvere to entwine.
Scarlet Pimpernel
Anagallis arvensis
Creeping on the ground, it is common everywhere from June to August. Pure red wild flowers are rare so if you see the flash of red in the grass you can be sure you have found the elusive pimpernel. If rain is imminent it closes its petals. For this reason it is also known as Shepherd's Weather Glass.